D0397250

has continued to grow at a staggering rate. What began as a cutting-edge practice of technology companies has become a necessity in nearly every industry. Between 2000 and 2015, U.S. multinational companies alone hired 4.3 million employees domestically versus 6.2 million employees overseas—that means millions of people who need digital technology if they are to communicate with the United States, not to mention the millions of domestic workers who work virtually from home over a distance of a few miles. McKinsey Global Institute predicts that the global labor workforce will reach 3.5 billion people by 2030. Remote work is increasingly here to stay. The future is in remote work.

None of these trends or predictions, however, accounted for a global pandemic that would require the wholesale migration of nearly entire companies to remote work in a matter of weeks. The remote work revolution, long in coming, was accelerated by the sudden and severe coronavirus outbreak. Chances are you are part of the massive transition that has forced companies to rapidly advance their digital footprint including cloud, storage, cybersecurity, and device and tool usages to accommodate their new virtual workforce. These changes have opened up a whole new scope of untold opportunities for people and organizations across the globe.

Now that companies have glimpsed the opportunities that remote work can provide, a portion will permanently maintain some remote days in their routines long-term. In a survey conducted in April 2020 by the Gartner group, 74 percent of 317 companies reported plans to indefinitely adopt more remote work post-COVID-19. Facebook, taking a gradual approach, expects to transition as much as half of its workforce to work from home within ten years. Stockholm-based fashion brand CDLP plans to restructure to locate 50 percent remote workers around the world. JPMorgan Chase, which saw its traders triple their productivity working from home, announced that the

INTRODUCTION

In the first few weeks of 2020, a microscopic agent turned the world's workforce into remote workers seemingly overnight. With the emergence of COVID-19, employees from China to Qatar, India to Australia, Brazil to Nigeria packed up their offices and set up new workspaces in their homes. Digital tools such as Zoom, Microsoft Teams, Google Chat, and Slack went from useful supplements to the primary enablers for daily interactions with coworkers.

These rapid changes were unprecedented, but the remote work format is not new. Domestic and global companies have had virtual work arrangements for nearly thirty years. Unsurprisingly, technology companies were the first to see the opportunities that remote work offers. The prominent technology company Cisco launched one of the first systematic remote work programs in Silicon Valley in 1993. Employees worked from home or kept flexible hours by using broadband technology to communicate with the main office from any remote location. Cisco reported saving $195 million in 2003, as well as an increase in employee productivity, both of which it attributed at least in part to its remote work arrangements. Also in the late 1990s, Sun Microsystems, still emerging from its startup status, instituted a voluntary virtual work program for 35 percent of its employees as part of its global expansion strategy. Within ten years, Sun saved half a billion dollars by reducing 15 percent of its real estate holdings (2.6 million square feet) in California while adopting geographically dispersed teams to be closer to their markets.

Since then, global teamwork—and thus the need to work remotely—

REMOTE WORK REVOLUTION

CONTENTS

To Lawrence, Gabe, and Daniel,
may your worlds always be without borders.

REMOTE WORK REVOLUTION. Copyright © 2021 by Tsedal Neeley. All rights reserved. Printed in the United States of America. No part of this book may be used or reproduced in any manner whatsoever without written permission except in the case of brief quotations embodied in critical articles and reviews. For information, address HarperCollins Publishers, 195 Broadway, New York, NY 10007.

HarperCollins books may be purchased for educational, business, or sales promotional use. For information, please email the Special Markets Department at SPsales@harpercollins.com.

FIRST EDITION

Designed by Bonni Leon-Berman

Library of Congress Cataloging-in-Publication Data
Names: Neeley, Tsedal, author.
Title: Remote work revolution: succeeding from anywhere / Tsedal Neeley.
Description: New York: Harper Business, 2021. | Includes bibliographical references and index. | Identifiers: LCCN 2020044119 (print) | LCCN 2020044120 (ebook) | ISBN 9780063068308 (hardcover) | ISBN 9780063068322 (ebook)
Subjects: LCSH: Telecommuting. | Organizational effectiveness.
Classification: LCC HD2336.3 .N44 2021 (print) | LCC HD2336.3 (ebook) | DDC 658.4/022—dc23
LC record available at https://lccn.loc.gov/2020044119
LC ebook record available at https://lccn.loc.gov/202004412021

21 22 23 24 25   LSC   10 9 8 7 6 5 4 3

REMOTE WORK REVOLUTION

Succeeding from Anywhere

TSEDAL NEELEY

HARPER
BUSINESS

An Imprint of HarperCollins*Publishers*

REMOTE WORK REVOLUTION

company is considering a permanently remote workforce, while UBS already expects as much as one-third of its employees to work remotely on a permanent basis. Groupe PSA, Europe's second-largest car manufacturer, announced a "new era of agility," in which its non-production staff will shift to remote work. The internet company Box expects more than 15 percent of its workforce to work remotely full-time after the pandemic. Similarly, Coinbase, a cryptocurrency exchange, declared it will become a "remote-first" company, estimating that 20–60 percent of the company will work remotely after restrictions are lifted, with more to follow over time. Nielsen Research in New York City will have its three thousand workers work from home most of the week. Nationwide Insurance, observing no loss in employee performance while avoiding operational costs during lockdown, will be transitioning employees at sixteen of its twenty locations into remote workers. Tata Consultancy Services announced that it plans to have about 75 percent of its workforce working remotely by 2025. Other Indian multinationals have also followed suit, as Infosys and HCL Technologies expect 35–50 percent or fully half of their employees to work remotely post-pandemic, respectively. The list goes on.

Rather than making plans to incorporate part-time or temporary remote work, Twitter and Square, both headed by CEO Jack Dorsey, made the bold move of giving their workforce the option to work from home "forever." Other companies, such as Slack and Shopify, have also responded to the call, announcing that they would extend indefinite remote work arrangements to most of their employees. An upstart company, Culdesac, took it a step further and declared the company was giving up its San Francisco office for a full-on remote work format with the hopes of creating opportunities to create a novel "homeless" nomadic culture. More are bound to follow.

As you may have discovered, there is no doubt that remote work has benefits. Commute times disappear. Operational costs get slashed. Bloated travel budgets are no longer imperative. Hiring and retaining employees without asking them to relocate from their home countries or domestic cities becomes conceivable, resolving global travel barriers. Astronomical real estate costs that exist in some locations have the potential to get reduced significantly, a welcome solution in an economic downturn. Societal ills like poverty gaps between rural and metropolitan areas might have the opportunity to close while simultaneously creating an untapped labor pool for companies. Gender gaps may shrink as organizations rethink their remote capacities for maternity leave. Gas emissions can decline, having measurable impact on environmental sustainability.

For workers and leaders around the world, however, untrained remote work isn't a panacea. In fact, you may have experienced some or all of the many challenges that are inherent with virtual arrangements. You are not alone if you feel isolated, out of sync, and out of sight. The more time we spend without regular in-person contact with coworkers, the more persistent and urgent questions about bonding, trusting, and alignment become. If your team finds that videoconferences lead to tech exhaustion, questions increase about how to choose the best digital tools to communicate. Or you may be one of the many for whom learning how to best structure tasks to optimize time and to avoid distractions at home is a priority. Agile teams need to transfer tightly coordinated work processes that rely on geographical proximity to a distributed setting. For leaders, how to keep employees motivated and consistently productive while monitoring progress from a distance is a source of concern. Because global teamwork by definition spans multiple geographies and cultures, questions abound about how to ensure that remote workers engage and collaborate effectively across borders. And above all, COVID-19 has brought into sharp relief that all leadership is global precisely because of the interconnected nature of the world; thus, inquiry about preparing for—or rapidly responding to—global events is part and parcel of a remote work revolution.

Remote Work Revolution provides evidence-based answers to those pressing concerns as well as practical guidance for how you can, together with team members, internalize and apply the best practices that matter the most. Teams and leaders who use this book will have the cumulative knowledge and skills necessary to break through routine norms and embody enduring behaviors that benefit yourselves, your groups, and your organizations as a whole. The book uses a wealth of lively stories to explain the problems that arise in remote work that most teams and leaders must navigate in order to rise to their organizations' highest levels. Each chapter draws from the work of leading experts in psychology, sociology, and technology, fields that are crucial to remote work success.

I have been deeply involved in the issues of remote work and global organizations for nearly two decades. As a professor at Harvard Business School, and before that, in my graduate work at Stanford University, I have researched, taught, consulted for, served on advisory boards of, and written case studies about thousands of distributed and global organizations in companies headquartered in France, Germany, Japan, and the United States, as well as in subsidiaries located in Australia, Brazil, Chile, China, France, Germany, India, Indonesia, Italy, Japan, Korea, Mexico, Russia, Singapore, Spain, Taiwan, Thailand, the United Kingdom, and the United States. In my work I have discovered that providing answers to questions alone, however, is never sufficient. Books and articles on remote work abound, yet questions never stop. Just providing information or answers on the topic doesn't help the key ideas to stick or behaviors to change for good, either. When people return to their daily demands, they easily fall back into old routines, and become frustrated and wonder why their teams don't fully cohere.

That's why, in both content and structure, this book is designed to engage directly with members and leaders of remote teams on practices that will help you bond and grow together. Through my consulting and advisory work, I have learned that the best way to internalize the values, norms, and behaviors for enduring success on distributed teams is to provide regular content that syncs naturally into work routines. Ideally, managers will deliver key content covered in this book to ensure that team members are laser-beam focused on success factors for remote work. As the insights build cumulatively over time, you and team members will expand your virtual teamwork capacity and help deliver results that were previously out of reach. Going through parts or the entire book together can provide a common vocabulary and set of practices that all members employ.

To effectively ingrain new insights from this book, *Remote Work Revolution* includes an action guide that has brief and varied activities for each chapter at the end of the book. Much as going to the gym exercises the body, the retrieval activities exercise memory, while enhancing team bonding in the process. With each retrieval exercise, you and your team members apply the information so it can begin to take root in your minds instead of disappearing after an initial pass. Retrievals are designed to ask readers to recall, describe, analyze, and apply best practices in a multistage process.

The cataclysmic and worldwide migration to remote work that COVID-19 engendered may have sped its completion, but this book had in fact been well under way. The accumulated insights and guidance have not been rushed into print. Nor are they a temporary fix. The behaviors and best practices—about trust, productivity, digital tools, leadership, and success—have taken years to develop. If adopting some of them seems challenging at first, know that you and your coworkers are laying essential and enduring groundwork. We will not remain a 100 percent remote world. Instead, we will see virtual, distributed, and global work become significant parts of work arrangements that expand our repertoire, skills, and performance, promising to make us and our organizations that much better.

CHAPTER 1

How Can We (Re)Launch to Thrive in Remote Work?

James sank into his home office chair as he listened to his client through the headset. "You destroyed my children's future," said Cliff, his voice laced with as much disappointment as anger. "I saved for years for this. How could you let this happen? I trusted you and I did everything right."

James had nothing to say. He worked for one of the fastest-growing residential real estate firms in the United States, and he knew that Cliff was right. He'd been so sure that he could help Cliff and his family fulfill their home ownership dream, but now, realizing that he'd betrayed Cliff's trust paralyzed him with regret and guilt. "I'm sorry," was all he could say. "I'm truly sorry."

Cliff, an exemplary client buying his first home, was the type of person that made James love his job. But his apology could not make up for the mistake his team had made. When the call ended, James sank deeper into his chair and tried to figure out what had gone wrong.

He remembered his first conversation with Cliff, which had taken place over the phone. Cliff said he'd spent his whole career making sure to save from his hard-earned income, even though it meant forgoing vacations. In the weeks that followed, James had been impressed at the watchful eye Cliff had kept on the costly and competitive California real estate market and at how focused and decisive he was in the search for just the right house, one with enough room for his wife and three children and in a neighborhood with good schools. He'd completed the application and verification forms as fast as they appeared. How elated Cliff had sounded when James gave him the news that he'd been approved for the home loan! Cliff wasn't one of those complaining clients that James knew too well. Even when the loan process began moving slower than promised, he was patient. James had been glad to be able to reassure Cliff. "Your interest rate is locked. We will push this through. Everything looks great. I'll call you next week to schedule the closing."

"I'm so excited," Cliff had said. "I can almost feel the keys in my hand."

Real estate is a volatile business. After that phone call, everything changed for James and the remote team he relied on to fulfill Cliff's loan request. A change in interest rates spiked a sudden volume of loans as people rushed to capitalize on available deals. James and his team were inundated with an increased number of interested customers. Unfortunately, they'd responded with a reactive mode to this sudden busyness.

A week passed, then two. Cliff called again, wanting to know how things were progressing. "The whole team is hard at work funding the loan and scheduling the close," James had said. He tried to sound reassuring. "I'll text you just as soon as we finalize the paperwork internally." He didn't mention to Cliff how long it had been since he'd talked to the team member handling the finer points of Cliff's loan or how busy they'd all been.

And then the call today. James's heart sank. Cliff had told him that his income had unexpectedly fallen. To avoid layoffs, his company had decided to impose a 25 percent salary reduction to everyone at his level. Cliff's voice shook with anger. "I don't have to tell you that I'm no longer going to make enough to qualify for the loan. Even with my excellent credit record. Last week, I would still have qualified! If you hadn't taken so long, I would be holding the keys in my hand by now!"

James wanted to blame the volatility of the real estate business for Cliff's lost opportunity. But the truth of the matter, he knew, was that volatility was what his remote team was unprepared to manage, slowing down the process so much that they ultimately failed a client. If only they'd been on the same page. If only they had taken the time to hold a meeting to devise a coordinated plan to meet the increased customer interest. Even a half day spent on reviewing and reorganizing work processes would have made a difference. If only they'd held a relaunch session.

A launch session (and periodic relaunches or reappraisals), which puts in place a clear group plan to meet the demands at hand, is *crucial* in remote work. Precisely because virtual workers are often distributed across many different geographical locations, work requires explicit planning. Like James and his team, those who are out of sight can fall out of sync at even the slightest bump in the road.

Holding reappraisal sessions may seem counterintuitive. In the thick of an overloaded work schedule, with deadlines whizzing by left and right, the concept of talking *about* teamwork instead of actually doing it may seem like an extravagance. Like James, many of us respond to limits on our time by immediately picking up the pace without the slightest pause for reflection. But this thinking could not be more misguided. In his decades of research, the pioneering expert on effective teamwork J. Richard Hackman (whom I will discuss in more depth later in the book) determined that actual day-to-day collaborative work is only the tip of the iceberg—10 percent, to be exact. With what he called the *60–30–10 rule*, Hackman concluded that 60 percent of team success depends on *prework*, or the way in which the team is designed; 30 percent depends on the initial launch; and only 10 percent depends on what happens when the actual day-to-day teamwork is under way.

Teams are always worse off without a proper launch session, no matter what. For a team to execute on its tasks—whether collocated, remote, or hybrid—it needs the right ingredients and the right preparation. This point may seem obvious, but it is very often overlooked for the reasons described above. While the "prework" determines what shape the team will take—its function, composition, design, etc.—and thus happens even before the team itself exists, the *launch* takes place at the moment the team comes together. As Hackman puts it, the team launch is what "breathes life" into the team by ensuring that every member understands and agrees on how they can work together most effectively. If teams skip this step, or brush past it in an effort to start work immediately, they often lose direction and falter down the road.

Team launches (and periodic relaunch sessions) drive performance throughout the team's journey. Relaunches are critical to keep remote teams cohesive, but even more so when teams transition to remote work, and especially by necessity as with COVID-19. Leaders need to be proactive about more, not less, periodic relaunches. The typical length for a launch is an hour or an hour and a half, and that time can be spread across two sessions. Every member needs to be present for an open discussion to share opinions and contribute perspectives on the best ways to work together as a team. When working remotely, launches should be video meetings where people can be as connected as digital technology allows.

This chapter will take you through the theory and practice of team launches by detailing the four essential elements of teamwork that each member must agree on.

1. Shared goals that make plain and clear the aims that the team is pursuing.

2. Shared understanding about each member's roles, functions, and constraints.
3. Shared understanding of available resources ranging from budgets to information.
4. Shared norms that map out how teammates will collaborate effectively.

Notice that each of these four domains begins with the same word: *shared*. That's because the fundamental goal of a launch session is alignment.

Relaunches are periodic appraisals of how the group is faring with the four key areas. I often joke that a relaunch is like a couple's date night because in both instances you revisit what's important and may check in on the present, past, and future to figure out what's working and what might need adjustments. As a general rule, teams should revisit their standing via a relaunch at least once per quarter. When people work remotely, I have found that relaunching every six to eight weeks to orient or reorient based on evolving dynamics is more important. During these occasions, virtual working groups and leaders acknowledge how each member is doing, figure out how to address concerns, and ultimately get everyone on the same track to achieve team goals.

In other words, relaunches are never a one-and-done event. Because work conditions are often dynamic, hitting the reset button once won't be enough. Periodic relaunches are important in good times but crucial in times of uncertainty, as James's story illustrates. The team might need to switch to a new mediating tool that calls for new norms of communication. The government might introduce new regulations or laws that affect people's work patterns, as we saw when millions were switched to working from home during the first months of the COVID-19 pandemic. Countries, markets, or entire industries might make a sudden shift that requires the team to reorient their goals. Periodic relaunches are the only structured mechanisms to give teams the ability to quickly pivot in a systematic way.

GETTING ALIGNED ON SHARED GOALS

Contrary to what many people believe, team alignment is not synonymous with agreement. In fact, *dis*agreement—often miscast as the enemy to cooperation—is a crucial part of refining ideas, identifying mistakes, and growing as a collective unit. The difference between successful and failed team alignment is thus not a matter of whether teammates disagree, but *what* they disagree about. Steve Jobs famously said, "It's okay to spend a lot of time arguing about which route to take to San Francisco when everyone wants to end up there, but a lot of time gets wasted in such arguments if one person wants to go to San Francisco and another secretly wants to go to San Diego" (1997). In other words, teams can disagree on the *how*—that is part of the dynamic process of teamwork—but before that process can even begin, teams must build a shared understanding of the goal, or the *what*. In Jobs's analogy, everyone must first be in agreement about the goal of traveling to San Francisco. Or bringing a specific product to market. Or growing a customer base. A team launch session is the opportunity to identify clear and specific team goals before taking any other steps forward.

To ensure there is shared agreement on the goals that the team has been mobilized to accomplish, the launch session must be a dialogue. As leaders and team members offer input, ask questions, pose concerns, and respond to others, they begin to understand and buy into the goals from their own perspectives. Leaders can make sure the conversation stays focused on the big picture. Quibbling over the details of *how* is a necessary conversation—but for a later date. Perhaps your team's goal is as simple as "delivering value to stakeholders in our industry." The only condition is that every team member agrees on and is fully committed to advancing it.

CONTRIBUTIONS AND CONSTRAINTS

Surprisingly, people are not always clear about where they fit on the team. A launch is an ideal opportunity for each member to explicitly articulate their individual roles and how they can contribute to the team goals. One member might volunteer previous experience on a similar project. Another member might acknowledge inexperience but express enthusiasm to learn. Someone else may show their aptitude for a specific skill that is needed to meet the team goal. As in a sports team, when members play different roles on the field, leaders may help identify each member's specific area of responsibility. Team members should understand everyone else's role as well as their own.

Clarifying individual team roles and each person's current responsibilities will also guide expectations about individual time and attention for collaboration. Remote team members often belong to multiple teams simultaneously. Multiple team membership—or at the very least, interdependence across teams—leads to varying or even conflicting expectations about how much time a team member devotes to each team. A team leader might assume that a colleague is prioritizing their team, when in actuality that person is focusing more on the work for another team. In fact, it is not uncommon for an individual to be involved in company work that isn't visible to teammates and managers. On a collocated team, someone's absence is an obvious fact. But in a remote format, there is no visual evidence of how team members spend their time. A launch that includes a frank discussion of such constraints enables the team to set expectations around how team-mates allocate their time to different commitments.

While misaligned expectations could make teams less effective, the opposite can also be true. In fact, the ability of each team member to candidly share their unique constraints and perspectives on how work should be done can be a strength. To capture the dynamic nature of multiple team membership for employees, relaunch sessions give team members the opportunity to discuss new assignments that they might have had added to their load. Awareness about how team members are coping with the added demands prepares the team to support each other, manage deadline expectations, and rebalance task loads ac-cordingly.

RECOGNIZING RESOURCES

One of the benefits of working in teams is the ability to draw on the distinct knowledge and skills of your teammates to help accomplish tasks. When team members are collocated in the same office, they can use each other's resources when collaborating face-to-face. But when distributed, opportunities for in-person interactions may be limited or nonexistent. Imagine two scenarios: In the first case, you've been working in the same office with your teammates for years. You're hud-dled around a conference table discussing the details of an important project. Because you know one another's strengths and weaknesses, it's easy to ask for or offer an opinion or specific piece of information. Now imagine that you're still discussing plans with your teammates, but instead of working in the same office, you're working remotely;

these days, you only meet your team members on videoconference calls or online chat rooms. If you've had personal experiences working together in online environments, you already know how difficult it can be to establish an interdependent dynamic that shares information and makes decisions.

On a material level, a launch session should identify resources pertaining to information, budget, technology, and internal or external networks that will help the team advance its goals. Generating a detailed list of every single item is not necessary, but the launch period is the time to reach a general consensus about the team's current resources, what it needs, and how to access them. Especially in a remote environment, this is the time to ensure that individual members of the team have the right technology and support to get work done. We can't assume that everyone has proper internet access, and some might need upgrades in devices or additional ones. Making sure that all employees have the proper financial support to outfit a working home office is imperative.

Relaunches are occasions for the team to reappraise available resources. For example, COVID-19 might have affected the team's budget and partnerships with other organizations. Individuals and leaders must make sure team members are aware of what is still within the team's tool kit as they press forward.

ESTABLISHING INTERACTION NORMS TOGETHER

Consider this scenario: A remote team of six excitedly downloads the latest chat applications on their smartphones. Even though they are distributed across five countries, now they chat anytime in a more informal way than email. One day, four members in the same time zone hold an impromptu conversation on the new app to discuss possible fixes of a bug on a software program. From there, the conversation jumps to an issue that the entire team is going to meet about the following day. The informal and ad hoc nature of the private conversation allows the four members to be candid with one another about their ideas and make significant headway on the tasks to be discussed at the scheduled meeting.

You might think that the team of four should get extra "points" for getting a head start, but on the contrary, the consequences do not bode well for the team's overall cohesion. At the meeting the next day, it becomes immediately clear to the other two members that they have missed something. References are made that they don't understand, and their questions go unanswered because the others are so far along in their dialogue. "Why are we being excluded?" they ask themselves. One doesn't want to bring it up explicitly for fear of seeming petty. The other had recently complained about meeting times and doesn't want to be pegged as the complainer. Even as they remind themselves that the exclusion was most likely unintentional, they feel a tinge of resentment toward their teammates—and a lingering worry that they will be excluded again in the future. Their resentment festers, and the team begins to splinter. I've seen this exact scenario, and many others like it, play out to the point that it shreds the team's cohesion.

What this team needed was a launch (or relaunch) session in which they discussed norms to guide their use of the new chat tool. The session would have given them the opportunity to understand and acknowledge that everyone must feel included on team progress if the group is to remain cohesive. They could have decided, for example, that if a few team members end up spontaneously chatting about a topic, they hit pause on the conversation until they notify the key players who are absent. The point is not so much the specifics of how a team chooses to communicate, but that the norms for doing so should be vetted before they are enacted.

Successful remote teams adhere to the group norms that they establish collectively. Norms are not rules. Rather, norms reflect a set of principles that guide interactions, decision-making, and problem-solving. Generating the norms together is essential; during the launch dialogue, members will learn about the issues that matter to their teammates. For example, if the majority of people on the team value punctuality, the launch can set an explicit norm about logging into meetings on time. This creates a standardized expectation for the whole team to follow. For the minority of teammates who are more casual about exact times, the established norm provides an incentive to show respect for teammates' preferences.

For remote teams that lack the opportunity for the informal daily interactions that happen in a shared office space, such as passing one another in the hallway or chatting by the coffee machine, norms defining virtual communication patterns are essential to filling the gap. Effective norms for communication have three primary functions:

- Outlining interaction and connection plans for all team members regardless of role or location
- Fostering psychological safety or the group's level of comfort in expressing individual concerns to one another about tasks and errors
- Keeping each remote team member connected so that no one feels professionally isolated

Plan Your Ongoing Communication

The most effective teams share one deceptively simple norm for communication: at meetings, each person talks and listens equally and makes an effort to address everyone (not just the team leader). After the meeting, people continue to work though relevant topics by resuming the conversation informally with other team members or searching for information that might help in the next discussion. In remote teams, for instance, based on a videoconference meeting with the entire team, you might shoot a message to a particular teammate on the internal social media tool: "Hi, great point you made about the project. It gives me a few ideas . . ." Your one-on-one chat might spur a brainstorm session to discuss in the next formal team-wide meeting.

Launch and relaunch sessions are opportunities to determine the best ways to conduct meetings and keep in touch with one another throughout the work process. For example, on one remote team I studied, using digital software to draw or write while discussing ideas (as might be done on a whiteboard when collocated) was the best way for team members to convey nuances to one another in a virtual format. Having a visual dimension, the team concluded, was the most efficient and pragmatic measure to achieve buy-in and mutual understanding. As one member put it, "For us, seeing is believing."

Teams also need to predetermine how to stay connected about work tasks. When must you let a teammate know that your delivery date will be later than scheduled? James's team suffered from a lack of coordination about who was doing what and when. James had periods of long silence when he had no idea what was happening with individual client work. Better visibility into Cliff's situation when their office became suddenly much busier would have saved everyone a lot of heartache.

Communication norms also determine etiquette for when to reach out, and when to follow up (if at all). For remote team members working from home, the boundaries between work and home can get blurry. Norms that strive to make the distinction as clear as possible— whether it's limiting correspondence to normal business hours or maintaining consistent expectations for punctuality and attendance in online meetings—help mitigate the confusion, exhaustion, and frustration that can arise when work and home life become entangled.

Make It Psychologically Safe for Conflicts and Mistakes

Collocated teams tend to argue more about work than distributed teams do. At first blush, this might appear like one less thing to worry about in remote work. But as anyone with experience on a remote team will tell you, all smiles and nods in a virtual meeting does *not* mean that everyone actually agrees with one another. Tension can exist without overt conflict, and it's much worse to keep that tension bottled up behind the screen than to hash it out with open dialogue. Task conflict is often a good thing—in chapter 7, I will discuss how to manage this process thoroughly. For now, it's important to understand that when people voice diverging or even opposing views, the dialogue often leads to more innovative and refined ideas.

Psychological safety, the condition that allows coworkers to take risks and admit mistakes without fear of reprisal or shame, is key to productive teamwork. My colleague Amy Edmondson, who pioneered the concept and has done extensive work on its effects, has found that if psychological safety is not present, people's fear of expressing dissent or uncertainty to colleagues—especially superiors—cripples team success. To counter these fears, leaders and their teams must actively foster an atmosphere that makes everyone feel safe speaking up and asking questions. When errors are out in the open, people are able to discuss how to reduce these errors in the future. The result is a team that is constantly learning, engaging, and improving.

Remote communication norms should lay the groundwork for a psychologically safe team. For example, a team launch session might establish a zero-tolerance policy for insulting language or map out a standardized process for moving forward on issues if teammates can't reach a consensus. Leaders may set the conditions for psychological safety by admitting their own mistakes and by expressly asking individuals to contribute thoughts and opinions.

Ensure That No Member Feels Isolated

Even if remote teams do a good job of promoting inclusion and psychological safety, the remote format is an inherently solitary experience for many. While research has provided ample evidence for the many benefits of remote work—wider geographical reach into different markets, more autonomy over one's office setup, and the list goes on—studies also make very clear that remote workers' feelings of professional isolation lead to reduced job performance and increased employee turnover. However, the effects of professional isolation on job performance decrease with some face-to-face interactions and access to more communication technologies, such as videoconferencing and dedicated voice over IP lines. Even the perception that coworkers are easily reachable helps allay the feeling of being alone.

Launch sessions are a time for teams to proactively set norms that make teammates accessible to one another while working apart. In this way, team norms must directly address how to mitigate feelings of isolation that may arise because of the physical distance among members. For example, the team could devise a plan that would break up solitary work by building in regularly scheduled or periodic face-to-face interactions. In the event that in-person contact is not possible, technology can be a worthy substitute. After all, physical presence is not necessarily the antidote to isolation—don't forget that people who sit together all day in an office without ever exchanging a word or a glance can also feel isolated from one another. Leaders can help convey the norm that isolation is a factor that the team will overcome together; this can go a long way in psychologically connecting members.

LEADERS NEED TO SHOW THEIR (RE)COMMITMENT

Launch sessions are also an opportunity for leaders to reinforce their commitment to the team. Consider Jennifer Reimert, the leader of a consulting team at Workhuman, a company that develops recognition and performance software for many of the world's largest organizations. As Reimert puts it, she is in the "thank you" business. Before starting at Workhuman, in the two decades she spent focusing on employee compensation, rewards, and benefits at a high-tech firm, she saw how the simple act of managers recognizing and thanking employees (and other peer-to-peer expressions of appreciation) served as an empowering force to boost engagement. In remote work especially, managers don't always witness the positive contributions that people make. Peers do. Recognitions that capture teammates' positive contributions create a culture of gratitude and positive reinforcement of the values that members espouse.

Reimert learned these lessons firsthand as a remote worker from the very start of her career. When she began at the high-tech firm, the job offer was three thousand miles away from her new home— her husband had been accepted to an MBA program on the East Coast, and they had recently relocated. After some discussion with leadership, they realized that her positioning in the Eastern Standard Time zone was perfect for the team, which had people in California, Oregon, the United Kingdom, and eventually Asia. By the time she began at Workhuman almost twenty years later, she had developed a set of principles for launching—and relaunching—remote teams. At the core of these principles was the belief that a strong team requires a leader who shows deep commitment to team members at every stage of the team's life.

In conjunction with team launch sessions, Reimert connects with her remote team members in one-on-one telephone sessions. She often has these conversations walking from room to room of her home, or, if the weather is nice, strolling on the sidewalk. Activity, she finds, keeps her focused on the person at the other end of the line. Her goal, as both a leader and a human being, is to listen to others, empathize, and respond accordingly. She begins the conversations sharing a bit about herself so that people can immediately feel a sense of comfort and familiarity. As they warm up, she asks for the person's honest feedback on the group launch session—what they feel optimistic and concerned about. She asks each person about personal interests, what they considered their strengths to be, where they want to improve, and what they want to get out of an experience on the team. Eventually, the conversation leads both Jean and the team member to an understanding of how the individual's interests, skills, and goals connect to the team's overall goals. Reimert finds this personal touch crucial for creating alignment in a group of people who work together without ever meeting in person.

As teammates get to know one another virtually through their shared work, Reimert makes a conscious effort to recognize each individual's contribution to the team. These small thank-you gestures go a long way to building cohesion. She also keeps her virtual door open for teammates to bring any concerns as they arise. Although she does her best to support and empathize, she also keeps in mind a key lesson that she says took years of perspective to learn: not everyone will be happy all of the time. In other words, don't fret that there will never be 100 percent satisfaction 100 percent of the time.

Reimert's approach illustrates a crucial leadership attribute for launching remote teams: leading by example. Her one-on-one sessions with teammates exemplified the communication patterns conducive to a psychologically safe and inclusive team culture. Teammates were compelled to respond in kind. In this way, the courage to show vulnerability at the outset *reinforces* a leader's role on the team instead of diminishing it.

When a team is in agreement in these four areas—goals, roles, resources, and norms—members become motivated and invested in meeting their team's goals.

Success from Anywhere: Launch and Relaunch

- **Set the compass.** Launch and relaunch sessions are an opportunity to set clear and precise aims for the team. Teammates work together better when they know that they all have the same goal in mind.
- **Talk about how to work together.** Set norms that guide communication patterns for an inclusive, psychologically safe, and connected team.
- **Know realities and fill the roles.** Be intentional about discussing how each member contributes to the team goals, what their internal and external constraints are, and where they can improve.
- **Find the resources you need.** Discuss the information, budget, tech, and networks that you and your teammates need to reach the team's goals. If you don't have access to these resources yet, discuss how to find them.
- **Show how you stay committed.** If leading launch sessions, show your appreciation for teammates by giving them your undivided attention, hearing their ideas and concerns, and responding with the resources available. Relaunches are a time to reinforce that commitment, especially in times of instability.

CHAPTER 2

How Can I Trust Colleagues I Barely See in Person?

Tara stared into her computer screen. She was gripped with anxiety. After having spent two days trying to identify the source of the bug in the software update, she finally admitted to herself that she had no idea how to find it. None of the other engineers on her small team had a solution, which meant she would need to find help elsewhere in the company—a multibillion-dollar tech company with more than 17,000 employees across thirty countries. Who would she ask? And if she could figure that out, the prospect of reaching out to a stranger horrified her. What if she was perceived as incompetent for failing with this task? She was relatively new at the company, and wanted to give off a good impression. Her mind swam amid countless question marks.

Then she had a sudden "aha" moment. She thought back to an all-company email that popped up in her inbox a few weeks ago announcing the launch of their private internal social media platform. The goal, as the email explained, was to promote knowledge sharing across geographically dispersed employees. "It's like Facebook for work," read a line above a link to register. Tara associated Facebook with her social life outside of the office, and the idea of blurring those boundaries gave her pause. But she needed a lifeline. So she opened the email and registered within minutes.

She eased smoothly into the platform's interface. Soon she was scrolling through fellow employees' pictures of pets and discussions about mountain climbing. People were definitely being "social," Tara observed. Then, a post about swimming piqued her interest. As an avid swimmer, she was excited to find something in common with another software developer on the platform. The developer was named Marisol, and her profile picture featured a woman with shoulder-length brown hair in her midthirties. Tara read a previous post in Marisol's feed. Another newly recruited engineer—similar in experience to Tara—had asked for advice on a programming issue, and Marisol had responded promptly and enthusiastically with clear guidance. Tara exhaled a sigh of relief. Although she had never met Marisol, she felt sufficiently confident that she could reach out to her without fear of being embarrassed or rejected. And that's what she did.

Put simply, Tara decided to trust Marisol. Social scientists define trust as the extent to which we are confident in, and willing to act on, the words, actions, and decisions of another. In other words, we trust people if what they say, do, and decide instills confidence.

TRUST IS NOT EVENLY DISTRIBUTED

When everyone works in one office building, even if not in close proximity, establishing trust in colleagues can be as easy as breathing—or as refilling your mug at the nearest coffee station. It's natural to strike up casual conversations with colleagues who work in different departments or in different teams. We gather all kinds of personal and professional details about who they are and how they comport themselves that make it easy to pass trust back and forth between each other. I call this trust-building process, which we think of as conventional trust, default trust.

But how do colleagues in remote work who seldom meet in person, if at all, discern that others are reliable? How do we develop concerns for coworkers' welfare when we work remotely so that we can feel reasonably comfortable interacting with one another? Collocated workers usually establish trust through credible, repeated interactions over time and shared contexts, yet this is difficult in remote teams where there are typically fewer in-person interactions and social cues. And what happens to long-established ties when they are tested by remote work over long periods of time? This is one of the questions that COVID-19 will force us all to confront, as we continue to spend more time alone in our home offices—away from the day-to-day, spontaneous, and informal interactions that create trust. How can you trust someone if you can't read gestures, body language, and facial expressions in periodic face-to-face meetings? How do you establish trust when you are separated by geography? How do we "read" trust in our coworkers when we're all relying on digital communication tools to do our work? How do we build new relationships with new team members?

What's more, trust is fragile. In most workplaces, trust is more likely to break when coworkers don't meet their responsibilities, withhold information, or form in-groups and out-groups. Managers who are perceived to "play favorites" or who implement what appear to be sudden and unnecessary layoffs are likely to lose the trust of remaining staff. Employees who consistently fail to do their best work may lose the trust of supervisors and coworkers. The problem is that once broken, trust is difficult to repair.

Although we may think of trust as a binary, one-size-fits-all experience, social scientists who have studied trust in the workplace construe trust as fairly nuanced and complex. You might think of trust as a palette, with different colors of trust to use for different circumstances.

The kind of trust Tara gave to Marisol is referred to by social scientists as *passable trust,* and it is an essential feature of remote teamwork. Passable trust is the minimum threshold of trust required to communicate with and to work with others. Another way to think about it: it's a sufficient level of confidence you have in others based on their words and actions. Passable trust is given to others through their observable behavior (whether in person or online or both). In this instance, Tara gained passable trust in Marisol by perusing another employee's interaction on the social media platform.

In addition to the passable trust that Tara felt in reaching out to a colleague "met" on the company's social media platform, social scientists have construed what's called *swift trust.* First identified in flight teams and law enforcement teams who were brought together in crisis situations, swift trust characterizes the high-level of trust that must be "swiftly" established by members in a team formed for a specific project or assignment who expect to be working together for a limited period of time. When swift trust is the norm, members decide to trust one another until proven otherwise. Recently I was a member of a team composed of faculty from across my university as part of a new dean's search advisory committee to work with the president and provost. In most cases we did not know each other well, yet we were tasked to handle a fairly sensitive matter. We had to swiftly decide to trust one another to keep all of our discussions confidential among ourselves. We had no other choice.

In this chapter you will learn more about these two types of trust, how they differ from default trust, why they are both essential in remote work, and the mechanisms that you and your coworkers can use to encourage them. You will also hear about how a financial services firm develops high-touch trust with clients. And because trust takes place in time, and is dynamic rather than static, it's helpful to think about what I call the *trusting curve*, a new way of understanding how trust works in remote teams.

THE TRUSTING CURVE

No doubt you're familiar with the term "learning curves." Originally conceived as a way to calculate the rate of improvement in performing a task as a function of time or cost (as in assembly line work), today we often talk about the learning curve as a way of measuring how long it takes to get better at a particular skill or task. People may move at different rates up a learning curve; for example, a gifted athlete who takes up a new sport may move faster up the learning curve than a previously sedentary person. Tasks, too, may require different amounts of time; for example, learning to write code may take longer than learning to work in a presentation template. People talk about learning curves as "high" and "low" or "shallow" and "steep." For our purposes, the important thing to understand is that a learning curve takes place in time. When graphed on an X-Y axis, the horizontal line is always "time."

In much the same way, we can conceptualize and graph trusting curves as taking place over time. In other words, the horizontal line of the graph is again "time," but the vertical line is "trust." In conventional terms, especially if face-to-face interactions are the norm, trust is a slow buildup over time; as time increases so does trust between teammates. But remote teams don't always have that luxury and require different pathways to developing trust, even if there are occasional face-to-face interactions. That's why the question in remote work should not be: Do I trust my colleagues? The question should be: How *much* do I need to trust them? As we look at the various types of trust suitable for virtual work, we can also understand also how they fit on the trusting curve.

TRUSTING THE HEAD AND THE HEART

Trust is the glue that binds a team together, drives performance, and enables collaboration and coordination, but you can't force trust. It is a judgment people must reach on their own. By trusting our colleagues, we are willing to be vulnerable to them when it comes to making sure that they will do their part in tasks or keep in confidence whatever we might confide in them. In teams, trust includes an expectation that people will act for the good of the group.

Two basic terms that help us think about how to choose from the nuanced palette of trust available when working together are *cognitive trust* and *emotional trust.*

Cognitive trust is grounded in the belief that your coworkers are reliable and dependable. Teams motivated by cognitive-based trust use their heads to consider their colleagues' qualification to do the task at hand; trust is usually formed over time, and confirmed (or disproven) over numerous experiences and interactions. For example, when you learn that a colleague has gained significant experience from a previous job or has graduated from an institution you respect, you begin to form cognitive trust. As you work on a project together, your cognitive trust will rise or fall depending on how consistently your colleague has behaved to demonstrate reliability over time.

By comparison, emotional trust is grounded in coworkers' care and concern for one another. Relationships built on emotional trust rely on positive feeling and emotional bonds, and they crop up most easily when team members share common values and mind-sets. If you consciously mentor a colleague or a group takes up a collection to give a coworker a gift, for example, that's driven by emotional trust. Relationships based on emotional trust are akin to friendships and involve the heart. They do not require more time to achieve, but they are more difficult to form among remote teams.

Passable trust is more dependent on cognitive trust, whereas swift trust is more dependent on both emotional and cognitive trust. Passable trust is necessary but not sufficient for most remote teams. It's useful and frequently used for communicating outside of teams and across organizations—it's the fuel that keeps organizations working—but because it doesn't become more intense or involve emotion it's not the special ingredient that makes a team, but especially a remote team, really gel. (see Figure 1 for Cognitive Trusting Curves and Figure 2 for Emotional Trusting Curve).

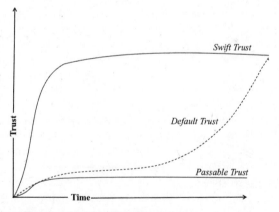

Figure 1: Three Types of Cognitive Trusting Curves

The trusting-curve graphs show that while an intense level of cognitive trust is typically reached fairly quickly with remote colleagues, it takes longer to reach emotional trust. Thus, remote colleagues are likely to work together with a relatively shallow level of emotional trust and a level of cognitive trust that is relatively intense. Note that while emotional trust takes longer to develop, it can eventually meet up and combine with cognitive trust; the two types are not mutually exclusive. One type isn't necessarily better than the other; what's important when designing or leading remote teams is to understand what kinds of trust exist and how they can increase collaboration and productivity when deployed well. How do you curate the right kind of trust for your team? What's essential?

In the next sections, let's look more closely at how these various trusting thresholds work and how you can achieve them. Understanding the different dimensions of trust and where you and your teammates may be located in the trusting curve will help to frame trust experiences as well as factor into your management and leadership of teams.

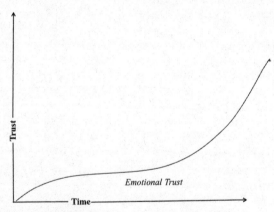

Figure 2: Emotional Trusting Curve

COGNITIVE PASSABLE TRUST

The passable trust that Tara developed when she reached out to Marisol was sufficient for her to get the help she needed. She didn't need to develop an emotional relationship with Marisol to do her job; passable trust was enough, and it didn't need to progress to something deeper or more intense. In six weeks Tara may reach out to Marisol with another question, and the passable trust they have will be sufficient and unchanged.

For virtual groups, who communicate largely via digital technology, without the benefit of a consistent common location to contextualize daily work habits, passable trust is especially useful, frequent, and common.

COGNITIVE SWIFT TRUST

Jerome thought he knew what it meant to be a team player. As a former nurse, he had worked with doctors and fellow nurses to save patients' lives in the ER department. Deadlines were literally a matter of life and death, and trust in one another was key to the success of the unit. But when he switched careers midlife and found himself in a new role working for an international medical device company, his notion of trust and team building was put to a test.

He and four others were required to collaborate on a marketing presentation for a new product. However, each of his teammates was located in a different country and the collaboration was completely virtual. No one had worked together previously. At the hospital, Jerome had been face-to-face with his colleagues every day, working together minute-by-minute to triage and treat patients. How could he rely on complete strangers who lived thousands of miles away? And through a screen, no less.

Jerome missed the first deadline. The initial exchanges among his teammates were all casual chats about hobbies and holiday plans—no discussion whatsoever about assigning roles or establishing team rules—so he assumed that they weren't serious about the work. But as he let his guard down and joined the conversations, he began to see how his teammates worked together with goodwill, curiosity, and diligence. Enrique, based in Chile, organized a schedule. Maria from Argentina and another team member—Sylvie from France—brainstormed ideas. Trude, who was based in the United States, drafted a summary list of their proposed options. They all reached a consensus on the most innovative idea to pursue.

So when Trude sent a panicked message with the subject line "URGENT!!! Idea rejected!" and explained their manager's objections, the dispirited team fell into a days-long tense back-and-forth. Jerome, no stranger to time pressure, saw the opportunity to help his teammates carry through. He made a strong case for pursuing another of the ideas on their original list. The team agreed.

For four days before the deadline, the team communicated digitally in real time. When someone had to log off, they would pass the work on to the others so that progress continued without interruption. After multiple drafts, they finished the presentation on time. They all thanked each other for a job well done and exchanged personal contact info. Although Jerome knew it was unlikely that they would stay in touch, he felt a deep sense of shared accomplishment from the experience.

Swift trust, the willingness of virtual team members to temporarily trust one another based on sufficient evidence of competence—whether it is work samples, pedigrees, or the way someone communicates with others in shared virtual spaces—is by far the most dominant formulation of trust in studies of virtual teamwork. While swift trust may be incomplete when compared to the default mode in which people get to know each other, it is sufficient to complete shared tasks.

Swift trust, crucial for remote workers who must immediately begin collaborating and coordinating, can be most challenging for people whose backgrounds or experiences value relationship building over time and easier for people who prioritize individualism and are task-oriented. Swift trust occurs in teams where members are connected through functional ties. Typically, swift trust is conferred at time zero, that is, at the moment we begin working together, and is then backfilled with evidence that accumulates over time as we work and interact with teammates to become durable. In other words, swift trust begins high and remains high because it is backfilled—with the caveat that it can drop if trust is broken.

KNOWLEDGE THAT PROMOTES TRUST

Developing trust in virtual teams follows many of the same conditions as in collocated teams. Leaders must set clear, superordinate goals and purposes, which team members need to understand and follow. Transparency, or sharing information freely, is important, as is effective communication, clearly identified tasks, reliability, and standardized internal processes. In virtual teams, every one of these conventional conditions must also include an awareness of how geographic divides and differences in people's lived contexts complicate trust. For example, in the beginning of a geographically distributed team's formation, when team members know the least about one another, and the sense of belonging to a group is least secure, is when individuals are most likely to adopt stereotypes that can lead to divisive subgroups (you will learn more about subgroups in chapter 7). To counter those tendencies, two additional mechanisms for promoting trust—direct knowledge and reflected knowledge—are especially relevant for remote teams.

Build Your Direct Knowledge About Team Members

Learning to trust and connect with one another virtually is even more effective if combined with *direct knowledge* about the personal characteristics and behavioral norms of far-flung virtual colleagues. When remote teamwork includes periodic face-to-face meetings we can make conscious efforts to learn about other people's lives rather than jumping immediately into a prescribed work agenda. Travel to a distant collaborators' site for a period of time to learn, for example, how one team member works under pressure and which individuals are likely to meet for working lunches, both examples of direct knowledge. When travel or physical proximity is not part of the work plan you can develop direct knowledge by taking the time to ask questions about teammates' own lives and work: "How is your home office set up going?" or "What do you usually do for lunch break?" The more context that virtual team members have about how one another works, the easier it is to trust them in their roles.

Consider the experience of Ben, who spent two weeks working side-by-side with his counterparts, Yee and Chi-Ming. During that time, Ben observes his colleagues in their context. Yee is calm under pressure, solicits counsel from key people when brainstorming, and goes to lunch with the same people daily at the second-floor cafeteria where people often hold working lunches. He notices the division of labor between Chi-Ming and Yee and when they draw Ben in. The direct knowledge that Ben gains about Yee and Chi-Ming provides him insight into their attitudes, behaviors, and motivations, and makes him more likely in the future to act on their words and actions—thereby fostering the development of trust among a mostly virtual team.

Develop Your Empathy Through Reflected Knowledge

Less obvious, but equally important for building trust among virtual teams is *reflected knowledge*, which is achieved by seeing your norms and behaviors through the lens of distant collaborators. Reflected knowledge is insight about how others see us and develops our empathy about how others experience us. The more that we feel understood by our colleagues, the easier it is to trust them. Consider the Indian engineer who complained that his counterparts in Germany were lazy because they took too long to respond to emails and seemed to work fewer hours than the Indian group. By the same token, German engineers complained that their Indian counterparts were always taking tea breaks. They didn't work hard. They were lazy.

In fact, the German engineers were accustomed to working sequentially on tasks with infrequent but carefully timed email communications and assumed the same of their Indian counterparts. Indian engineers did go often to the tearoom, often in pairs—to mentor, share knowledge, and problem solve. Had the two groups understood the processes of how their colleagues got things done, they would have been less likely to complain and more likely to trust the competence and motivation of their teammates.

Reflected knowledge allows each sub-team to learn the inaccuracy of their perceptions. In this case, both sets of engineers would understand that differing work practices, rather than laziness, were at the root of their frustration and distrust. Reflected knowledge would enable each group to better understand and adjust their perceptions. If the German engineers had the ability to see their work practices through the lens of their Indian colleagues they could reflect on the relative isolation and highly scheduled manner in which they worked and appreciate the highly collaborative nature of the Indian office. Likewise, the Indian engineers' reflected knowledge through the lens of German colleagues would allow them to reflect on the relative haphazard organizational practices in which they worked and appreciate the structured, planful nature of the German office. Understanding the norms of one's own site better enhances empathy, closeness, and trust felt for colleagues with differing norms.

We develop reflected knowledge remotely by paying deep and sensitive attention to the way our teammates work: how they communicate through email or video, the times of day that they sign on and off of shared virtual spaces, whether or not they tend to respond to messages outside of normal work hours, etc. Perhaps one teammate responds enthusiastically to an email at 9 p.m., and another responds the next day with a discernable tone of stress about being pinged late the night before. When you discover that norms aren't aligned, these observations help you adapt your own behaviors. And when they are aligned, reflected knowledge gives you the feeling of confidence that your colleagues understand you more deeply.

To facilitate the exchange of direct and reflected knowledge among virtual team members, leaders must proactively create a group culture for virtual interactions not explicitly related to work tasks. Social platforms that enable daily informal communication or virtual coffee/ tea chats that rotate among different team members or allotting time before and after virtual conference calls for nonwork chatting can work. Each team should find the easiest way to get together for these conversations. What matters the most is that everyone understands that the purpose of these interactions is not for team members to advance work tasks, but to get to know one another as individuals: asking questions about one another's interests outside of work, daily routines, preferences, workspaces, etc. As team members learn about their teammates through these conversations, they gain direct knowledge about their teammates' situations and perspectives, and as they learn about themselves from the new vantage point of their teammates' perspectives, they gain reflected knowledge.

EMOTIONAL TRUST

How do we develop emotional trust with people? One of the most powerful ways is by self-disclosure, or the process of making yourself known to others. For more than fifty years, self-disclosure has been widely studied across a variety of interpersonal contexts, including friendships, romances, and therapeutic relations. The way to enhance trust between people is for all parties to self-disclose, which increases a general sense of closeness and likability.

Self-disclosing to your teams has to be explicit, intentional, and voluntary. Self-disclosure occurs in what you say during meetings, write in emails or chats, and post as pictures or videos on the appropriate social media. It is particularly important for remote workers because visually apparent social cues and other observable information to build connection with others are elusive or nonexistent. Here are elements of self-disclosure that matter to receivers:

- Depth: level of intimacy conveyed
- Breadth: amount of information disclosed
- Duration: length of exchange
- Reciprocity: whether the disclosure is one-sided or an exchange
- Truthfulness: how "authentic" the information appears
- Attribution: if the information is uniquely intended for the recipient
- Descriptive vs. Evaluative: e.g., "I had dinner" vs. "I like Habesha food"
- Personal vs. Relational: e.g., "*I* like eating seafood" vs. "I like eating seafood *with you*"

What this means is that for emotional closeness to develop, we have to share a bit of ourselves in the casual conversation that may occur at the beginning or end of group meetings or in the course of digital communications with individual coworkers. "I can't meet at that time because I have to take my car to the mechanic." "I would have sent this sooner but I am struggling with my technology." "Our new client is from Connecticut—hey, that's where I grew up!" "I saw the picture you posted of Johannesburg. My family lived in South Africa for a year." The more you learn about someone, the more you will probably like them and the closer you feel. Without such sharing, especially in remote work, you end up with a one-dimensional transactional relationship that is only about the task at hand. Unlike in person, where the idle time you spend with your coworkers inevitably leads to serendipitous discoveries about one another—like how a colleague always makes a cappuccino on Fridays at four o'clock sharp—in the remote format, you have to make a point of sharing these kinds of quirks and habits.

Of course, self-disclosure also requires that we make judgments about the boundaries of what is and is not acceptable in specific contexts and what degree of personal information we feel comfortable disclosing. You might not want to share, for example, all the gory details of your recent surgery with your marketing team, but it would probably be appropriate information to share in a telehealth meeting with your doctor. While you want to be authentic in how you present yourself to others, certain comments that are commonly offensive (for example, sexist) are never acceptable.

BUILDING TRUST WITH CLIENTS FROM AFAR

As any leader knows, building trust internally—with peers, bosses, and direct reports—is critical. The same holds for external partners, particularly customers. How do we build trusting relationships with our customers and our external partners when the usual mechanisms for establishing emotional and cognitive trust—office visits, business meals, conferences—are no longer feasible?

One leader who has figured out ways to use digital media to effectively build trusting relationships with remote customers is John (pseudonym). His team aims to help their clients manage their liquid assets of $5 million and more. In their advisory work, they have to tailor investment strategies to meet unique clients' needs and interests. At the heart of this work is building emotional trust with clients.

Traditionally, building trust and establishing relationships with high-end clients has been done in person. Increasingly, however, the traditional in-person, "high-touch" methods are becoming less viable. Even before the COVID-19 pandemic eliminated virtually any opportunity for in-person contact, budget cuts and the increasing cost of airfare and other expenditures had limited clients' face-to-face contact to two or three times a year, which was insufficient to build and maintain trust. The group had to adopt virtual strategies, which, on the one hand, are more convenient, but on the other, require more creativity, time, and intentionality. Here John is building trust that is neither swift nor passable, and therefore is a testament that default cognitive and lasting emotional trust can be nurtured virtually.

John and his team understood that increasing frequency of valuable contact was going to be necessary. Such contact or touch-points can be created using the range of digital tools that are available, including social media, videoconferencing, and email. The goal is to be able to create opportunities for a higher number of interactions in a diverse range of ways. The second necessity was looking the part, or finding ways to make a virtual conversation feel as in-person as possible: this means dressing to the occasion—whether more formal or more casual—and finding the appropriate lighting so that your face is as clear and communicative as possible. Another crucial aspect of building such trust is concision and precision: unlike an in-person meeting, where one can give an extensive presentation, when interacting virtually you have to identify the most essential details and communicate them clearly in only a few lines. For John, this can be as simple as sifting through a newsletter, comparing the insights to the contents of the client's portfolio, condensing the findings into three lines, and immediately following up with a call. It can also, though, take on a more creative side: taking a cue from other industries, he and his team created short, colorful videos—in the same vein as the many "life hack" how-to popular videos—that briefly lay out the details of a new product in far less time and far more engaging ways than a traditional presentation. All of the above methods serve to build the client's cognitive trust, reinforcing the client's impression that John and his team are sufficiently reliable, credible, experienced, and knowledgeable to perform the job of managing millions of dollars.

John also found ways to have emotional touch points. For example, his team arranged a special birthday surprise virtually for a client. They ordered an arrangement from a florist who notified his team that they were about to deliver the flowers. John called the client at the point when the delivery appeared. When the client's doorbell rang, he excused himself to answer the door. Seeing the client's face in real time and to be present with him as the gift arrived strengthened their relationship in a unique yet emotional way. In another instance, a teammate sent a client masks at the height of the COVID-19 pandemic. This was particularly meaningful because his client was worried about finding masks for himself and his family. That gesture of concern increased the client's emotional trust; he responded by immediately sending over the funds for a product that had been under discussion.

John and his team also enlisted their professional network in fresh ways to create more social contact with their clients. For example, they created private "interest groups" on social media tailored to their clients. Those interested in wine are invited to a virtual wine tasting, in which each person is sent a selection of wines and invited to a video call, where they are joined by a professional sommelier and the members of their working team. Those interested in tennis or golf, meanwhile, are sent short, personalized video shoots with professional players. While not explicitly focused on selling, these personalized, highly tailored interactions are key to developing, from afar, a rich, trusting relationship that can climb up the trusting curve, which in turn pays dividends in working relationships later down the road.

Trust is the glue that binds virtual groups and assures work success. There will be times when default, conventional notions of trust that rely on credible, repeated interactions over time and shared contexts will be necessary for your work. Other times it's necessary that you just give your trust unless proven otherwise. Because trust is dynamic rather than static, you can use the trusting curve much like you would a compass, to ascertain where you are in the trust process—high and fast or low and slow—and where you might want to go. Discerning the magnitude and intensity of trust you need in a given situation has to happen with little face-to-face contact. The trusting curve in remote work is a tool that allows you to determine what you need and how long it will take for you to get there.

Success from Anywhere: Building Trust in Remote Teams

- **Use just enough.** An incomplete or imperfect trust that is sufficient to get information or work done is crucial to figure out in remote work. Observe, learn, and determine what information would be sufficient to discern that someone's actions or words are dependable for the shared work you need to accomplish.

- **Assume the best.** If necessary, you can quickly confer trust to team members sufficient to fulfill a shared task. Check out the information you need to determine coworkers' competence and give bounded trust as you accumulate information on whether you continue to confer trust or not.
- **Gain direct knowledge** to better understand the context in which team members' work is fodder to enhance relationship trust.
- **Study your own reflection.** Developing an empathic lens that lets you see how others experience you and your actions yields powerful information that can help you to cultivate meaningful trusting relationships.
- **Share yourself.** Much harder to establish without the benefit of time and mutual closeness, emotional trust conveys the feelings of mutual care and concern between coworkers based in shared positive emotional bonds. You get there by opening up to your team members so that they know who and how you are. Sharing autobiographical insights will help you get closer to your team members, thereby cultivating emotional trust.
- **Create new pathways.** Be attentive to your clients' needs and create virtual experiences that promote cognitive and emotional trust with them. You will need both. Digital tools can uniquely create meaningful experiences with them that will show them you care and that you are reliable.

CHAPTER 3

Can My Team Really Be Productive Remotely?

If you're like most people working on remote teams you are probably concerned about productivity. How do you measure productivity? How do you keep track of work? What if people get distracted or plain lazy away from the office? Netflix, playful pets, errand runs, or unchecked socializing might tempt people away from working. Even the best-intentioned might not be up to the physical and psychological challenges of working consistently from home.

Working remotely, whether full-time or in a hybrid arrangement, may raise concerns that your team understands that you are working on task, responsible, and committed. You may struggle with staying connected to the rest of your team or ensuring that your home can support the focus and concentration that you need to get the job done. Can you maintain enough self-discipline and self-direction? Or do you find yourself working around the clock, concerned about the toll that high productivity may take on connections with family and life apart from the job?

Managers worry about their remote team's ability to meet their organizational goals, including the fact that they are accountable for them. Without the ability to see teams in action firsthand, managers may worry about a worst-case scenario. Although a massive shift of a team to remote work certainly serves up its own twist, the truth is that most managers have limited power over employee productivity even when collocated. Teams don't always turn in reports by agreed-upon deadlines. New software breaks down. Customers are unhappy with service representatives. Unless you're a nineteenth-century factory boss sitting in a glass office above workers assembling widgets, having ultimate control of what your employees do or don't do is as far gone as the industrial era. Yet fears about managing remote teams have led many companies to install monitoring technology to keep remote employees productive from afar.

In this chapter we will first look at how surveillance technologies and tracking tools tend to backfire. Then we'll consider the larger issue of what makes teams productive by looking at the conclusions reached by the late influential sociologist and pioneering expert of teams J. Richard Hackman, who made it his life's work to figure out what *conditions* make for successful and therefore productive teams. Remote work, in one form or another, has been around for decades, which means that we've been able to gather plenty of data about its productivity—data that I discuss to demonstrate that in fact there is plenty of good news. What do people need to work productively when away from the office? While they typically value the autonomy and flexibility remote work affords, they may struggle with feeling connected to their team, setting boundaries between their work and home lives, and home conditions that challenge focus and concentration. At the end of the chapter you will find advice for creating the conditions that make remote workers productive.

SURVEILLANCE FOR PRODUCTIVITY

Consider the shock of a twenty-five-year-old employee of an e-commerce company when she opened an email from her manager asking her to install a piece of software that would track her keyboard strokes and the websites she visited on her own personal computer. Her jaw dropped even further when she read the rest of the email: in addition to the software, she was to download a GPS tracker on her personal phone. The measures were intended to ensure the company's productivity by trailing employee work behaviors all day.

An employee at another company described the shame and anxiety she felt when her company began using a digital apparatus that would take pictures of her at the computer every ten minutes to discourage idling around when working remotely. The apparatus also monitored the duration of her breaks and would display a pop-up message with a one-minute warning before she should resume working if she didn't want her daily log of hours to pause. As an hourly employee, these pauses cut into her earnings. The looming threat of the pop-up constantly preoccupied her—even if she walked away from her computer for a bathroom break or took a phone call that wasn't explicitly work related.

At a translation agency in Australia, managers can see every window open on their contractors' desktop computers at every minute of the day. Every move of their mouse cursors is scrutinized. Check-in emails flood their inboxes with expectations of immediate replies. Ironically, these draconian measures didn't exist when everyone shared a physical space. The fears companies have that workers will go rogue when unsupervised arise from the need to advance goals without actually observing firsthand the daily grind to get there.

Suppliers of these monitoring tools refer to them as "awareness technologies." One Connecticut-based company's business tripled when COVID-19 drove millions of people home. The sheer presence of their tools, the company argues, effectively curbs people's tendency to neglect their professional responsibilities if left unchecked. Euphemisms aside, the subtext is clear: if big brother is not watching, employees will slack off. The head of a social media marketing company seemed to agree with this premise. Once he (literally) lost sight of employees after they began working from home, he promptly installed digital monitoring devices in an attempt to obviate the sudden uncertainty that arose from not being able to see what was going on and assuage concerns about the consequent loss of productivity. Nonetheless, privacy advocates bristle at the proliferation—and potential permanence—of digital surveillance in the lives of workers, even if monitoring suppliers laud the tools as a helpful deterrent and managers may find comfort in the opportunity to gather data on employee productivity.

Not all tracking tools for remote workers are framed as mechanisms to police behaviors. Some managers attempt to re-create the constant companionship that employees enjoyed in a collocated workspace by requiring that video cameras and microphones stay activated throughout the remote workday. The idea is that the audiovisual presence of teammates—even if it is limited to a window within a computer screen—could cut through the isolation that is inherent to remote work and allow workers to engage in spontaneous interactions should the mood hit.

Whether installed explicitly as a monitor for productivity or framed more innocuously as a passive facilitator for constant connection, employees despise surveillance tools. The experience makes them feel self-conscious to the point of heightened anxiety and demoralized to the point of losing loyalty to their employer. Many tolerate the intrusion only because they fear job loss if they push back, especially when the economy is not favorable. Those who can afford to leave their companies often do. An analysis by Accenture found that employees became highly stressed and felt disempowered under the gaze of monitoring tools. A survey conducted by Deloitte found that millennials had intentions to leave companies that they perceived as emphasizing profits over people's well-being. In fact, the study found that the tools were unsettling even to those who were meant to benefit from the gazing, so to speak: a staggering 70 percent of the surveyed C-suite executives were ill at ease about the effective use of surveillance data.

Leaders must recognize the risks associated with digital supervision. Despite what may well be the best of intentions, digital surveillance *by definition* conveys a lack of trust between employers and employees—especially if these tools are an attempt to establish control after a sudden shift into remote work. When you signal mistrust in employees, you are eradicating the bedrock of effective teamwork. What good are "awareness technologies"—or any attempt to enhance productivity, for that matter—if the most basic conditions for a productive team don't exist?

ASSESSING TEAM PRODUCTIVITY

It's impossible to talk about teams and their productivity without understanding the work of J. Richard Hackman. No one understood team dynamics like he did. Known for climbing into the cockpits of airplanes or searching for truth about teams in the most unexpected places, for four decades he examined groups from every conceivable context—C-suite teams of major corporations, orchestras, Central Intelligence analytical teams, hospital care teams, flight crews, and more. At Harvard, where he was on the faculty for many years, his belief in developing people was legendary. His presence filled a room, his deep voice commanded attention, and for some reason, every time I saw him—whether leading a seminar or chatting one-on-one—he seemed to be entangled in an intellectual debate, fiercely arguing his points with unending empirical evidence. After his passing I was surprised to learn that he was only a little over six feet because his towering presence made him look several inches taller. He has deeply influenced my thinking as he has at least two generations of scholars and teams.

Hackman established that team performance can be assessed by a specific set of standards. One of his enduring contributions includes three criteria for establishing successful outcomes for teams that are applicable across the board, regardless of industry or context: 1) delivering *results*, or achieving expected goals; 2) facilitating *individual growth*, or a sense of personal development and well-being; and 3) building *team cohesion*, or ensuring that the team is operating as one unit. As I will describe later, these criteria are crucial to explain the failures that monitoring in the name of productivity poses for remote workers and their organizations.

Delivering results is one of the fundamental questions you probably ask when assessing productivity. For client-oriented projects, effective teams successfully deliver relevant goods or services. For internally oriented projects, effective teams fulfill their necessary functions: strategy teams successfully develop strategy, ops teams successfully manage operations, tech teams successfully implement technologies, and so on. Of course, there isn't a one-size-fits-all description that defines exactly what constitutes "successfully" delivering a project or fulfilling a set function. A product team might bring its product to market under budget and within a deadline—meeting the expectations of leadership and stakeholders—but sacrifice quality and disappoint its customers. Individual teams must define expected goals for themselves.

A second fundamental measure of team performance involves individual experiences. In successful teams, people learn and feel that other members care about their well-being or *individual growth* as a function of being on the team. Teamwork, therefore, offers opportunities for each person to expand their knowledge, acquire new skills, and be exposed to new perspectives. Even if these opportunities don't have a direct effect on the team's measurable results, individual growth often leads to increased job satisfaction that in turn enhances the team's productivity. In the absence of this criterion, people can develop negative sentiments. Nearly everyone has at one time or another worked in a situation where they feel that they are not developing and that their emotions are unmet within the team; consequently, their engagement is likely to diminish. For teams to be effective, each member must feel optimistic about their individual role, what they can offer the team, and what the team can offer them.

The final measure, *team cohesion*, assesses the extent to which team members operate as one unit. Learning how to work together as a group—rather than as individuals working in silos—is what creates a cohesive team. The key ingredient to this learning process is social connection: team members must feel sufficiently connected if they are to collaborate efficiently as a singular group. Very often, this process takes time. Through the experience of working together, team members can develop strategies that enhance coordination, develop collective skills, and maximize the team's efficiency.

REMOTE WORK INCREASES PRODUCTIVITY

Here's the good news: the fears that inform some managers' gut reaction to use surveillance tools are unfounded. Studies show that remote work does not pose a threat to productivity; in fact, remote work actually *increases* it. Managers who adopt policing strategies miss a central fact about productivity, namely, that it comes from the trifecta of team results, individual growth, and team cohesion. As I will illustrate in the rest of this chapter, the features of remote work align with this trifecta in multiple ways—when it comes to the connection between team results and individual growth, for example, working from home affords employees more flexibility in arranging schedules, gives them more autonomy over their work environment (no more thermostat wars), and saves time on commutes. Later in the chapter I will share the key practices that are necessary to increase productivity when working remotely.

Let's first briefly review our understanding of productivity.

Companies and scholars have been studying the efficacy of modern remote work for nearly three decades. By modern, I mean virtual professional engagements that are enabled by digital tools (not the distributed work of the late 1600s, in which London-based merchants sailed across the Atlantic to coordinate with colonists on the North American coast, just for clarification). As mentioned in the Introduction, makers of technology products were first to experiment with modern remote teams. When Cisco launched a remote work program in the Silicon Valley area in 1993, more than 90 percent of their employees participated in the grand experiment to work from anywhere. People had the freedom to choose their desired place to be on the job—coffee shops and kitchen tables, as well as the office, were acceptable options. It didn't take long for Cisco to reap the financial benefits from the savings on hefty real estate upkeep with fewer bodies at the official workplace. Cisco credited an uptick in the focus and dedication that remote work inspired as reportedly saving $195 million from a rise in productivity within a decade.

Another technology company, Sun Microsystems (a decade before it was acquired by Oracle in 2009), had built a diverse workforce that needed to collaborate across multiple time zones and functions within their workplace locations. Along with the unique needs of a distributed team structure, employees also expressed a desire for more flexible work arrangements. That's why, starting in 1995, Sun's top management began brainstorming possible options, ultimately designing and launching a remote work program that they called "Open Work." The leaders concluded that they needed to enable employees to work from anywhere, anytime, using any technology. At the time, this was considered a fairly unusual idea.

Similar to Cisco's initiative, Open Work combined a three-pronged approach that involved technologies, tools, and support processes, in that order. Remember that in 1995 connection to reach the internet was challenging. Mobile phones were not widely used; Bluetooth and the cloud did not yet exist. Open Work therefore offered a suite of enabling technologies referred to as "mobility with security," in which people moved between work sites and had consistent mobile access to personal computer sessions. The second novel idea for the time involved access to workplaces on a day-to-day basis in spaces that included the Sun campus, a drop-in office, a hoteling site, or a client site. Third, employees could work from home or a Sun workspace when an alternative space was needed. To make Open Work accessible, Sun provided their mobile employees monthly allowances to pay for the cost of internet, telephone, and hardware. Today, community shared workplaces are a thriving business and a way of life—shortly before the stay-at-home orders were issued in Massachusetts, my local Staples renovated to include a handsome shared workspace that featured a free space for local community meetings—but twenty-five years ago a company such as Sun found it necessary to design training to help people adapt to the new way of working.

Once the alternative work arrangement was launched, about a third of Sun employees decided to opt in, which meant that they wouldn't use their assigned buildings on a typical workday. Apparently the arrangement was popular, because that number nearly doubled within ten years, with roughly 60 percent participating in the program. Sun also benefited by reducing its real estate holdings by more than 15 percent, resulting in savings of nearly half a billion dollars.

These growing trends piqued the interest of management scholars who wanted to figure out whether the lived experience of remote workers was better or worse than that of their on-site colleagues. One study hypothesized that those who worked away from the office would show an increase in measures of productivity because they no longer had to endure the time and stress involved with commutes and that the convenient work arrangements would heighten people's sense of job satisfaction, provided they had good relationships with their teammates. The study's hypothesis proved correct. Remote workers loved their logistical conveniences. No more anxiety about getting to early morning meetings on time. No more staring at long traffic lights or dodging impatient drivers recklessly switching lanes in a traffic jam. No more back pain from a cramped driver's seat or a crowded bus. The study found that remote employees who could get to work in the time it took to walk from their kitchen to their desk had 30 percent higher productivity than their commuting colleagues.

Was this a U.S. phenomenon, or would the same productivity gains emerge in a different cultural context? How might an experiment in remote work unfold in a Chinese company, where different cultural norms exist to define distinctions between individual and collective needs in organizations? A group of economists teamed up to assess the benefits that work-from-home practices can have on performance and productivity at China's largest travel agency, Ctrip. One of the study authors, James Liang, was a cofounder of the company, and thus had a vested interest in this question. Interestingly, when they asked 996 employees from the Shanghai call center if they were interested in working from home, about half expressed interest, but only 249 met the company qualifications of having at least a six-month tenure, broadband technology, and a private workspace to do their job at home. The scholars randomly chose to study 125 employees who worked from home, while the other half continued to go to the office. Nothing else changed. Each group continued their responsibilities to care for customers for the next nine months.

What did they discover nine months later? Not only was working from home desirable, but comparing the time that each group spent logged on to take calls, they discovered that the remote workers increased their productivity by 13 percent compared to their colleagues. When they looked at turnover they found a 50 percent drop compared to the group that continued to work at the office. Over the course of the experiment, Ctrip improved total productivity by 20–30 percent and saved about $2,000 a year for each employee that worked remotely—largely from reduction in office space, increased performance, and reduced turnover. Encouraged by the results, the firm extended the option to work remotely to everyone. Performance doubled to 22 percent for those who accepted the offer.

The examples I have provided so far clearly show the productivity and financial benefits of remote work in for-profit sectors, where performance is closely measured from a number of criteria. Suppose the employees worked for the federal government, and therefore didn't have to worry about quarterly performance. Scholar Raj Choudhury—along with his partners Cirrus Foroughi and Barbara Larson—approached the United States Patent and Trademark Office (USPTO) to see if examiners working remotely were more productive in a government context.

The USPTO is a federal government agency headquartered in Alexandria, Virginia, and spread across eleven buildings. The agency's chief imperative is to fulfill an article in the U.S. Constitution to "promote the Progress of Science and useful Arts, by securing for limited Times to Authors and Inventors the exclusive Right to their respective Writings and Discoveries."

When citizens have a unique idea that they want to protect, they have to go to the USPTO to work with one of the patent examiners assigned to their case. From my own personal experience with getting a patent for my software simulation on global collaboration, I can tell you that a patent application can take years to get through. Examiners are skilled workers, but they are not in a hurry to close applications. Lengthy, detailed, and highly technical forms must be carefully reviewed for approval along several channels, providing multiple opportunities for stalling or bottleneck.

Choudhury and his collaborators had the opportunity to look at two remote programs at the USPTO. The first was a work-from-anywhere format with employees who lived approximately fifty miles from the office. The second was a work-from-home option of up to four days per week. Employees needed to have had at least two years of satisfactory performance to participate. Roughly eight hundred patent examiners met the criteria and joined the study. When the work-from-anywhere program was compared to the work-from-home option, they found a 4.4 percent increase in output by the group who had the freedom to choose where to spend their workday. Once again, we see a clear boost in productivity. But more precisely, we also see how much people value their freedom to choose their remote arrangements. The desire for autonomy at work is a consistent and striking pattern that we see, and one for which remote work is particularly well suited.

REMOTE WORKERS NEED AUTONOMY

The hallmark of remote work success is the ability to self-direct and capitalize on the gift of managing your own work processes. Hackman's insight on the importance of individual growth extends to remote workers' need to choose where and how they work. In fact, a through line across decades of studies into remote work identifies autonomy as pivotal to job satisfaction and performance. By autonomy, I mean the ability to self-govern. In remote work, this translates to flexibility in the timing and location of work. With the exception of periods that call for coordinated efforts with teammates, having control over where, when, and how you work matters a great deal—and for good reason. It signals trust and reliability (which in turn boosts self-confidence), it allows ownership over projects (which in turn boosts personal investment in the project's success), and it allows the tailoring of your workday according to individual schedules (which in turn makes for more efficiency).

This last benefit—flexibility of scheduling—is particularly invaluable for remote workers who have to negotiate the demands of work and family at the same time and is often touted as one of its more appealing benefits. The monitoring that I described at the start of the chapter has the exact opposite effect that autonomy affords: by signaling that employees are untrustworthy and unreliable, surveillance discounts their agency on team projects, and forces workdays into a fixed schedule. It is the epitome of overcorrection—a straitjacket that protects against an unlikely worst-case scenario by eliminating any potential whatsoever for movement.

Does autonomy really shape work experiences and individual outcomes? Employees were surveyed as part of an ongoing study at a large telecommunication company. Of the study participants, 83 were remote workers and 144 were not. The remote workers reported significantly more autonomy, more cross-discipline collaborative projects, and more career advancement prospects, and significantly less time spent on what's called "strain-based" work-family conflict than their peers who worked at the office. The study suggested that the increased flexibility and control over work arrangements might have been the reason for the decrease in work-family conflicts. What's more, despite feeling like they received less career support—likely because they spent little time with managers—remote workers didn't report any barriers to their career mobility opportunities.

With autonomy, or the capacity to self-govern, also comes the capacity to make commitments. Typically, if people feel more committed to something—an organization, a cause, an idea—they will work harder to achieve their goals. Commitment is an important proxy for employee retention, and retention is good for productivity because companies don't have to rehire and retrain but instead can depend on a stable of experienced workers. Studies confirm that when people have the opportunity to work virtually and the flexibility to arrange the job tasks there is an increase in commitment to their companies and in performance, and a decreased likelihood for attrition. However, feelings of exhaustion weakened these results. Not surprising, exhaustion can erase the sense of control that is paramount to pretty much any job satisfaction. Likewise, exhaustion from the volume of task activities must also eat away at the flexibility that remote workers prize.

A team of scholars wondered if issues of autonomy would be different if remote workers were part of a large and distributed workforce across the United States rather than a smaller team. They used at least three methods to learn about the employees' behaviors and productivity, including surveys, interviews, and managers' performance ratings. Their study involved approximately one thousand employees over time. Some employees worked remotely, others did not. Consistent with patterns of previous studies, the scholars found that the employees who perceived greater psychological job control had significantly lower turnover intentions, family-work conflict, and depression.

WORK CONDITIONS HAVE TO BE CONDUCIVE

Sean is a software engineer who works remotely for a video game company. Ever since he could remember, Sean has loved two things: coding and video games. Sean's team at work loved his uncanny ability to resolve any technical puzzle. He would pursue any problem relentlessly. He liked working with his team. As a young adult, he had always been able to block out the world and sit for hours, lost in his creative programming. He would sometimes forget to eat breakfast and lunch. Laser-beam focused on the screen, he was eventually able to impeccably handle thousands of lines of code. Life changed for Sean when he happily married his college sweetheart; the couple had two children, a daughter and a son. Sean loved his family, coding, and video games—in that order. Yet he started to feel increasingly dissatisfied with the conditions in his home. The sharp focus that he relied on to help his team and fulfill his own responsibility was no more. His wife complained that his mind was always on work, even at mealtimes, and he grew irritated at what felt like constant interruptions about seemingly trivial domestic concerns. Space was an issue. Noise was constant. His family's economics wouldn't allow a way to change his home conditions. For the first time in his professional career, he considered abandoning his remote arrangement. The boundaries between home and life were impossibly blurring.

Sean wasn't the only one who felt troubled trying to balance work and life at home. Millions of people around the world experienced the blurring of work and home-life when stay-at-home orders were issued during the COVID-19 pandemic. This can be disruptive if we are accustomed to drawing a line between our home lives, which tend to be relatively private to most of our coworkers, at least in certain cultures, and our professional sphere. Of course, we hope and assume that our employers care about our well-being, but the extent to which they can poke around in our lives is necessarily limited. My colleague Lakshmi Ramarajan has always held that people's identities are multifaceted. Working professionals may simultaneously define themselves as, for example, technical experts, task force contributors, members of a national group, and parents. As you may well know, multiple identities can bring advantages—life is richer, the world is larger—but they can also bring struggles for people trying to reconcile or jump back and forth between different sets of behaviors and values in and out their professional circles.

In addition to other identities, remote workers who work from home must by definition toggle between a work and a home life. Although for many parents, remote work can allow more flexibility for family time and child care—for example, school drop-offs, pickups, homework, and meals—some studies have found that remote work arrangements can cause more family-to-work conflict among males than females.

Home conditions—workspace, technology infrastructure, privacy, and people—can determine whether people will thrive remotely or not. The flexibility in the timing and location of work that virtual arrangements can bring is only beneficial if you feel your home situation complements rather than interferes with your job. Of course, household size can either weaken or intensify the issue. People with smaller households experience less interference than those with larger households. Adequate home office space can also determine satisfaction. Cramped apartments or makeshift spaces in the bedroom corners can be downright uncomfortable and hinder concentration. Sleeping in the same bed or eating off the same table at which you work contributes to the feeling that you are "living at work," rather than "working from home." For Sean, home conditions changed when small children came on the scene, but for many other people, roommates or multi-generational households can be the home condition that decreases job satisfaction.

Home conditions play into well-being. Overall, studies have found that the flexibility in timing and increased opportunity for work-family balance that working from home can bring are beneficial to well-being, and thus worker satisfaction and productivity. However, for others, the blurring of boundaries between a professional and personal self, or home conditions that interfere with the focus demanded by sustained work, can cause conflict and unease. Working from home, in other words, depends on what kind of home it is and who else is in it.

TEAMS NEED COHESION

The inherent difference between remote work and collocated work is also the most salient. We are no longer nearby the very people who have by turns delighted us, supported us, and irritated us (or vice versa); we no longer see them hunched over computers or walking down the hall; we no longer hear their voices echoing from the conference room or their shared laughs emanating from the coffee corner. By definition, to be "remote" is to be distant, unreachable, disconnected. A closely knit remote team sounds like a contradiction in terms.

Yet productive and satisfying working relationships are not dependent on physical proximity. Whether members are remote or collocated, team cohesion is a matter of working collaboratively and efficiently together. Team members feel cognitive and emotional connections with one another and unite around a common goal. They understand where they are on the trusting arc to help communicate regularly and coordinate tasks clearly. They rely on, trust, and appreciate one another and can learn from one another's strengths and weaknesses. They face conflicts head-on and work together toward resolution. A common location is not a prerequisite for a cohesive team. Researchers have found that workers can collaborate productively on a remote team with as little as 10 percent of their time spent in face-to-face interactions.

For remote workers, team cohesion depends on two interrelated factors: the frequency of interactions with other team members, and the quality of relationships that these interactions form. More important than collocation is the extent to which people feel *included* in the group: whether they feel recognized, engaged, and up to date on the team's progress. A survey in 2008 of a large, high-tech corporation measured feelings of professional isolation among the large percentage of its of employees working from home. Adapting the format of a long-established and widely used survey called the UCLA Loneliness Scale, the researchers asked respondents to read a series of statements and rate the extent to which these statements reflected their experience on a scale from 1 to 5. For example, statements read, "I feel left out on activities and meetings that could enhance my career," "I feel out of the loop," and "I miss face-to-face contact with coworkers." When they compared the survey results with productivity metrics, the authors of the study found that "professional isolation among remote workers was negatively associated with job performance."

The use of a measurement for *loneliness* to explain "professional isolation" is telling. In recent years, research has identified loneliness as a serious public health issue—equivalent to smoking fifteen cigarettes a day. The "cure," as this research notes, is meaningful relationships. What can we learn about professional isolation by examining how the UCLA Loneliness Scale defined this universal condition? Of the twenty items on the survey, not one mentions physical proximity. In fact, one question asks respondents whether they feel that "people are around me but not with me," implying that a person can feel lonely even in the presence of others.

In other words, professional isolation is a cognitive and emotional experience, *not* a physical position. Where one is located is not directly relevant to how one feels. Team members could sit beside one another every day and still be strangers. The answer to professional isolation, then, is developing cognitive and emotional connection with one another—regardless of the team's physical format. When those connections are strong, then the team is cohesive. And when the team is cohesive, it is productive. In fact a cohesive remote team—with all its inherent time- and money-saving advantages—has the capacity to be even more productive than its brick-and-mortar counterpart.

WHEN REMOTE WORK FAILS

Are some jobs better suited to remote work than others? When this question was posed to 273 employees who worked from home in sales, marketing, accounting, and engineering, the researchers found that highly complex jobs that did not require social support were more conducive to remote work than collocated work. This study also found that low-complexity jobs that did not require much interactive collaboration, such as call centers, were more productive when working from home. Even for workers whose jobs were more interactive, the researchers found no negative correlations between remote work and job performance. In other words, remote work doesn't significantly hurt job performance in any type of work. For some job features, performance is better with more extensive virtual work, and in others, the impact is neutral. Another study designed to capture the lived experience and consequences of working from home compared to the office found that employees did better on creative problem-solving tasks when working from home. If we leave out the high-touch jobs such as hair salons or tattoo parlors, many jobs thrive in a remote format—especially those that require deep problem-solving and undistracted concentration. Software engineers, graphic designers, editors, writers, or other knowledge workers who can do most of their work at a computer fit this category.

Success from Anywhere: Be Productive Virtually

- **Focus on process and not outcome when assessing productivity.** Equip your teams with the tools and resources they need and assume that they will have the insights into how best to achieve their work goals. Managers have to take to heart, as Ernest Hemingway famously remarks: "The best way to find out if you can trust somebody is to trust them."
- **Lean into the inherent flexibility** of the remote format. Instead of monitoring team members obsessively, encourage their autonomy. They will gain confidence, agency, and efficiency. The result is a more productive team.
- **Provide support for optimal working conditions** because they are pivotal and might require financial resources from your budget. Ask remote workers what they need to create the best work conditions for them, wherever that might be. Whenever possible, assist workers with resources and planning to ensure that they are comfortable with their work situations.
- **Emphasize team goals and identity.** Without a home base that has the company's name and brand scrawled above the entrance, remote teams need explicit reminders about their purpose. It is the leader's

job to keep remote team members aligned with one another on a shared mission, and show each person how they individually contribute. When team members feel included and purposeful, the team is cohesive. And when the team is cohesive, the team is productive on a level that collocated teams can't reach.

CHAPTER 4

How Should I Use Digital Tools in Remote Work?

On February 7, 2011, Thierry Breton, CEO of global information technology giant Atos, announced at a press conference that he would ban internal email. At the time, the company had over 74,000 employees. This was not a whimsical or impulsive decision. By the time he landed at Atos in 2008 (for a turnaround leadership job that he would pass with flying colors), Breton had been contemplating the efficacy of technology and its transformative nature for decades. He'd founded a software company in his early twenties. He'd written a novel, *Softwar*, in which the plot revolved around a computer virus released as cyberwarfare between countries—way back in the 1980s—which sold over two million copies.

What appeared to be Breton's sudden erasure of email was his radical response to what he deemed as the unnecessary volume of emails people receive—"email pollution," as he calls it—which gets in the way of working together. He also worried about how the onslaught of email in people's inboxes was causing them to work extra hours to respond. "We are producing data on a massive scale that is fast polluting our working environments and also encroaching into our personal lives," he said at the time of the ban, and went on to declare, "We are taking action now to reverse this trend, just as organizations took measures to reduce environmental pollution after the industrial revolution."

Internal emails were replaced by internal social networks, instant messaging systems, and collaborative tools. While Atos employees did not reach Breton's original goal of eliminating *all* internal emails within eighteen months, his bold plan dramatically decreased their use while increasing the use of digital collaboration tools at the company. The company culture shifted to embrace more instant modes of communication by using real-time calls over the internet (voice over IP) and videoconferencing. These modalities allowed employees to communicate in real time, also referred to as synchronously. In addition, the system that Breton implemented could easily show people's status on the network—whether they were on or off. This ongoing indication of who was present spurred people to initiate online conversations with coworkers, which in turn could spontaneously create a team interaction on the fly as people invited others or as more colleagues joined ongoing conversations. Ultimately, employees held online meetings, often using videoconferencing, from their local machines.

Although Breton was radical in his approach, he fully understood that as the leader of a distributed global organization, he needed to intentionally create the conditions for everyone to connect or collaborate despite the physical distances that separated them. He also understood that it's the leader's job to decide the desired communication culture and then choose the tools to achieve that for a remote workforce. Email never went away altogether at Atos, but to this day, the employees are adept at creating teamwork spaces or intentionally selecting the right media for their purposes. Breton has since been tapped by Emmanuel Macron, the president of France, to assume the role of European Commissioner to, among other things, help 511 million Europeans transform digitally.

Remote workers at every level routinely make choices about which technology-mediated tools will most effectively allow them to accomplish their work and at the same time enhance their relationships with coworkers. This chapter will address questions about how you should convey a message, whether to follow up an initial message through a different medium, or how to signal the importance of a message unobtrusively. When are written modes, like email or internal social media tools—modes that have permanence—warranted? What about real-time video or voice-dependent communications? Is it effective to send an email that sits in the recipient's inbox and serves as a visual reminder of an outstanding task? What is the best medium for collaboration in a group? What's the best way to convey information when you know people are drowning in it all day long? How do you maintain connection and continuity when you seldom work in the same place, if ever? How can you avoid tech exhaustion?

TECH EXHAUSTION

Let me begin by getting the issue of tech exhaustion out of the way. Complaints of cognitive overload, headaches, and even the slurring of words are often accompanied with complaints about going from one videoconference to the next. Tech exhaustion happens when we treat work communication activities in the virtual world in the same way that we do in the physical word, yet don't add the constraints that we do in the latter. For example, if we have consecutive in-person meetings, we always add transition time between meetings. This is partly because in-person interactions, which usually require some kind of travel time from point A to point B, even if only down the hall, don't allow for these tight adjacent meetings. One or two meetings may occur in close succession, but it doesn't happen with every single meeting every single day.

Exhausted remote professionals often schedule their meetings with one ending and another one starting immediately after. What's more, without building in time to process post-meeting and organize a to-do list that might have come from the meeting, it's easy for work to accumulate unnecessarily. Just because digital tools allow us to fully pack our calendar doesn't mean we should. It is, therefore, crucial to create a period of transition between meetings.

Equally true: just because we have videoconferencing available, doesn't mean we should be on video calls constantly. Don't get me wrong, video calls have many beneficial characteristics. Communication tools such as email, phone, videoconferencing, instant messaging, and social media are important to use in a way that fits the occasion, especially since they are not neutral conduits; rather, they shape social dynamics that influence the advancement of work goals. To excel in remote work, it is critical to understand how to choose the right digital tools to ensure that teams are using them productively to thrive in a distributed process.

The selection of appropriate digital tools for distributed communication harks back to the 1970s. The good news is that we know a lot about how digital tools seem to touch all of us and how they can be the most vexing for remote work in the absence of their deliberate use. I will describe the key problems—mutual knowledge and social presence—and solutions that are important to understand when making technology decisions for remote work. Rather than listing a limited, one-to-one correspondence, this chapter will give you a vocabulary and framework to inform your choices about which digital tools are appropriate to use when and under what circumstances.

For organizations and leaders designing remote work options, it isn't just a matter of choosing which suite of technical tools to acquire; it's a matter of understanding that different tools support different goals and come with distinct benefits and limitations. Some tools favor autonomous and asynchronous activities; others reinforce collaboration and real-time discussion. Some tools increase immediacy and intimacy; others are designed to formalize processes and policies. Email, text messages, videoconference, phone, and social media platforms are among the most popular in the plethora of available digital tools that are available are in common usage. Understanding the differences in type and characteristics and becoming intentional in your selection of these digital tools will make your teams more effective and increase team cohesion and job satisfaction.

But first, you need to understand the unique challenges that remote work presents for your workforce and for you as a leader. Whether all of your workers are remote or you are implementing a hybrid model where some workers are collocated and others are not, it falls upon leaders to determine the desired communication culture you want to create. Through my own research and those of other social scientists, I have identified at least five additional conundrums—beyond tech exhaustion—you'll want to resolve in order to answer this question:

- Mutual Awareness
- Social Presence
- Rich vs. Lean Media
- Productive Redundancy
- Cultural Differences

THE MUTUAL KNOWLEDGE PROBLEM

Shared assumptions and understandings are a requisite for effective communication. In a virtual world, sharing assumptions or holding common ground while being out of each other's sight defines a classic problem because in the simplest instances, we need extensive common ground to successfully interpret the situational context and respond appropriately to one another. Social scientists call this the *mutual knowledge problem*. If you agree to "meet Jenny at that café down the street for coffee after the conference call," then you're relying on the fact that you both know and understand relevant background assumptions, such as the name of the café, where it's located, and when you ought to be there. Similarly, a project team needs to share common ground about the project specifications—how to use the appropriate tools (for example, spreadsheets or cash flow discounting)—and have a shared understanding of what the output needs to look like to satisfy stakeholders. Successfully navigating obstacles and delivering results requires alignment. However, that's easier said than done. Failing to share common ground or misinterpreting assumptions can compromise project outcomes.

Why do remote work goals fall prey to the mutual knowledge problem? One of the most influential investigations into this issue looked at people who collaborated remotely across the United States, Canada, Australia, and Portugal for seven weeks on a group task. The group completed a multipart exercise that involved generating business ideas, writing a business plan, and creating a presentation or web page. During this collaboration, they generated 1,649 emails, numerous chat logs, and project outputs, all of which were analyzed to find out where teams fell in the mutual knowledge problem. The study found examples of several different types of failure.

Participants might fail to provide common ground by not contextualizing their work; for example, if they did not let teammates on one project know they were busy with other projects to explain their apparent under-involvement. Email behaviors could fail to provide common ground in a number of ways. Participants might practice uneven distribution by not sending emails to all team members, with the result that some members might be more "in the loop" than others. Emails might address several topics, so the importance of any given topic was not emphasized (what researchers called "underappreciating salience"), causing snafus in coordination and prioritization. Even something as seemingly benign as how often participants checked their emails—multiple times per day versus multiple times per week—made a difference in how fast information was accessed. Confusion also arose around what it meant if a participant remained "silent" in group electronic communications; some interpreted the silence to mean "I agree," while others thought it meant "I disagree," and still others thought silence was a neutral stance meaning nothing at all. Taken together, these blurry communication strategies led to unequal assumptions among team members that in turn created problems in alignment and productivity.

In addition to analyzing modes of failure related to mutual knowledge, the study found that when participants are distant, they have trouble fully recognizing and appreciating the circumstances of their collaborators, so if there is absence of information, they are more likely to attribute failure personally rather than to any other possibilities— which, in turn, made the search for constructive remedies tougher. For example, in the same way that people had a hard time dealing with silence in a real-time conversation, they also had a hard time when people didn't respond to email fast enough. People were more likely to interpret silence or what they considered lags in response time as personal failures on their part or personal affronts from the other party.

THE SOCIAL PRESENCE PROBLEM

To state the obvious, one of the challenges with remote work is the fact that we do not meet in person, face-to-face. Digital communication, in one form or another, is an attempt to remedy that challenge either by duplicating as much as possible what we achieve in face-to-face communication or by offering an alternative form of communication with benefits unachievable in person. But why is face-to-face communication so powerful? And what exactly is lost when we communicate virtually?

One way to frame the problem is to look at what social scientists call *social presence*. Face-to-face contact is considered the gold standard for social presence. But when in-person interactions are not available, we can turn to social presence to define the degree to which specific media convey social cues via voices or facial expressions to the extent that perceivers could understand the communicators' thoughts and feelings.

Two key concepts of social presence are *intimacy* and *immediacy*. Intimacy captures the feelings of interpersonal closeness that two people experience when interacting with one another. It is influenced by factors such as eye contact, smiling, body language, and topics of conversation of varying sensitivity. Thus, a digital medium through which people can see each other's faces in real time has a greater sense of intimacy than one without this capability. Immediacy refers to the psychological distance or feeling of mental or emotional connection a person places between themselves and the recipient. This can be conveyed both verbally and nonverbally, through physical distance or closeness, by what people are wearing (for example, formal or informal dress), and by facial expressions during a conversation. The technology used will obviously determine how much any of those characteristics can be seen and felt by people engaged in communication. Interestingly, immediacy can change even if social presence does not: for example, while the degree of social presence may remain the same between two people speaking remotely by phone, the immediacy may change if one of the speakers' attitude or tone suddenly changes from, for example, warm and open to harsh and critical.

Both intimacy and immediacy are governed by two additional aspects of social presence: efficiency and nonverbal communication. In this case, efficiency relates to the medium that a communicator deems the most effective to get a message across to an audience. While face-to-face interactions, for example, have the highest degree of social presence, in some instances, such as when there is a high degree of confrontation or interpersonal tension, another medium with less social presence might be preferable and therefore more efficient. Nonverbal communication, meanwhile, refers to the extent that a digital medium can show the same details that in-person interactions afford. People can communicate with the least ambiguity and provide the most information via nonverbal communications such as body language, eye contact, posture, and physical distance. Of course, people can try to control their nonverbal behavior consciously or subconsciously: we all know someone with a nonexpressive "poker face" whose feelings are difficult to decipher, or have had the experience of seeing someone appear positive despite hearing bad news.

So what does this all mean? It means that it is important to consider social presence and whether your media of choice is conveying your "warmth," "personability," or "realness." Visual and audio media can do so to differing extents; although audio media will not convey overt nonverbal cues or subtler insights, such as the apparent distance or authenticity of the people we are talking to, we can become skilled at hearing these things conveyed by tone and volume of voice. Ultimately, the media to use for communication is dependent on what we are trying to communicate, and that includes the degree of social presence we want to try to achieve. In remote work, digital media choice is dependent on what is appropriate for specific goals.

LEANER TO RICHER MEDIA

If you talk to any technology-enabled communication expert about how people should select their media to fit their needs, they will start talking about rich or lean media.

Rich media are those that convey greater amounts of information, including social cues and social presence, which enhance understanding across a wide swath of situations, even those that are ambiguous, whereas lean media are those that convey less information, fewer social cues, less social presence, and have relatively limited communication. Both lean and rich media are important and exist on a continuum. The richer media are more effective in situations with higher ambiguity, higher equivocality, and less clarity, while leaner media will be more effective in situations that are more straightforward (See Figure 1).

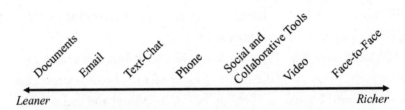

Figure 1: Examples of lean versus rich media

As you will notice, leaner media tend to be asynchronous while richer media tend to be synchronous. When thinking about what types of work activities were best suited for different levels of leanness/richness and synchronicity, researchers determined that communication is made up of two primary processes they call *conveyance* and *convergence*. Conveyance describes the transmission of new information from one person to another; for example, that a new shipment of a certain number of goods is expected on the morning of October 15. The receiver of the conveyed information might then need time to check the shipment's inventory against an original order of goods. Lean, asynchronous media would be suitable for this work activity. Convergence, on the other hand, describes communication in which individuals must discuss and interpret information to come to an agreement. Discussion about how to best use the shipment of goods upon arrival would require back-and-forth dialogue in a richer, synchronous media.

However, not all work activities can be assigned to solely to media that is lean, rich, asynchronous, or synchronous. Much depends on the circumstances. Sometimes richer media is called for to coordinate who among a group will be responsible for a large set of competing tasks, while other times a lean, asynchronous poll can coordinate the best time to schedule a meeting among a group of people. (See Table 1 for a breakdown of the types of work activities that match media characteristics in most circumstances.)

	Rich	Lean
Synchronous	Complex coordination Discussion Collaboration Team building	Routine coordination Information exchange
Asynchronous	Content development Team selection	Information exchange Simple coordination Complex information processing

Table 1: Work Activities and Digital Media Characteristics

Other researchers took this work a step further. They noted that although both a phone call and instant messaging are synchronous (or mostly synchronous) media, they do not provide the same level of effectiveness in all circumstances. Someone can easily send an instant message that will be transmitted simultaneously to many people; a phone call carries voice and sound, usually to one person or a relatively small group. By characterizing technology by specific capabilities, they identified five that mattered the most: the speed at which a medium can deliver a message to intended recipients, the number of people a medium can simultaneously reach, the variety of expressions a medium conveys (physical, visual, and verbal information), the extent to which the medium enables the sender to rehearse or fine-tune a message before sending it, and the extent to which the medium enables a message to be captured and made permanent to reexamine, reprocess, or repeat. These capabilities, mapped out in Table 2, further detail how we can think about the characteristics and effectiveness of certain media when choosing what to use when and under what circumstances.

	Delivery Speed	Number of Recipients	Variety of Expression	Fine-tune Potential	Message Permanence
Face-to-Face	High	Medium	Few to Many	Low	Low
Video-conference	High	Medium	Some to Many	Low	Low to Medium
Conference Call	High	Low to Medium	Few to Some	Low	Low
Shared Folder System	Medium to High	Medium	Few	High	High
Social and Collaborative Tools	Medium to High	Medium	Some to Many	Medium	Medium to High
Instant Messaging	Medium to High	Low to Medium	Few to Some	Medium	Medium
Email	Low to Medium	High	Few to Some	High	Medium to High
Documents	Low	High	Few to Some	High	Medium to High

Table 2: Comparison of Selected Media and Their Capabilities

Because face-to-face meetings are the richest form of communication, you might think that the goal in using digital communication is always to prefer the richest mode. While it's true that rich communication is often desirable for team effectiveness, it's not always the best choice. What matters more than richness or leanness is the *relationship* of team members to each other and the given communication goals. This is especially true when it comes to negotiations and group decision-making. Teams with generally positive relationships, such as those who may have friendships that extend beyond the office, benefit less than those from richer technology such as video. Because they already know and feel positively about one another, leaner communication methods such as email seem adequate for reaching agreement. Indeed, defaulting to richer forms of media on the assumption that "more is more," especially when a group is already tightly knit, may be one cause of tech exhaustion that so many employees reported experiencing when COVID-19 work-from-home orders first went into effect in March 2020. On the other hand, teams whose members have neutral relationships—for example, teams formed randomly by lottery or location—*do* achieve better outcomes when they communicate via richer media types, perhaps because they need more information about how the others think and act. Most surprisingly, teams who have a preexisting negative relationship to communicating with one another—for example, a history of disagreement or antagonism—actually fare *worse* when asked to negotiate and make decisions using rich technologies. When hostility is a factor, lean technologies can buffer or deflect potential unproductive conflict. In other words, when making decisions about the best communication technology to use, consider the dynamics and history of your teams as much as, if not more than, you consider the technology itself.

The purpose for which the technology will be used is also an important determinant. Researchers have found that teams working on "nonroutine" tasks such as drafting a report benefit from technology that uses "task knowledge awareness" so they can understand "who is doing what" to track, for example, who is responsible for writing which section of the report, and various due dates for delivery and revision. Lean, asynchronous communications work best for record keeping and scheduling. By contrast, when teams engage in activities that involve multiple languages or time zones, technologies that increase presence awareness (feeling present and interactive) such as video calls enhance performance. Put simply, different technologies affect team goals and performance outcomes.

Performance outcomes and mappings such as that of digital media capabilities in Table 2 are very helpful to guide media choices for remote workers, but they are incomplete because people seldom only use a single medium to communicate about one thing. People will use a mix of media to communicate with the same people. In fact, I have found in my own work that savvy media users mix and match their digital tools strategically.

REDUNDANT COMMUNICATION

Our instincts about good communication might suggest that being redundant is to be avoided for the sake of efficiency. But it turns out that social tools that increase and reinforce redundancy are not only useful but often essential for virtual teams. Maybe you feel fortunate because you work on a remote team that effectively prohibits people from showing up at your desk to inform you of the same thing over and over again; nevertheless, you've almost certainly been on the receiving end of redundant communications if someone has mentioned the same thing to you more than once. My colleagues and I conducted a study to examine the nuances of redundant communications that use multiple media by observing project managers at six companies.

As part of our study, I observed an early morning meeting in which Greg, a project manager, had informed a team of fifteen people that they would soon be transitioning to a new product development process. One person after another had raised various objections. It would be too time consuming to keep a closer record of activities. Why was the quality assurance manager being switched to another team? The time to verifiable product was too short. Greg had talked through each of the concerns with admirable patience and skill, and by the end of the meeting, everyone had reluctantly agreed to follow the new process. However, by 11:15 that same morning I observed Greg spend twenty minutes carefully crafting a twenty-line, two-paragraph follow-up email to the team in which he essentially used the same words as he had in the meeting about the new process. I watched as he changed the wording of the subject line three times until he came up with: "Design for Excellence Review" and ended with, "Thank you for helping to facilitate our progress and to ensure the process meets our guidelines."

Why, I asked, couldn't he just send a brief greeting and recap of the meeting to accompany the documents he needed to attach, which included a questionnaire that team members had to fill out to move the transition to the next step? Greg explained that despite the team's agreement to cooperate, because of the initial opposition they'd presented he felt he needed to do more to persuade them. He emphasized the urgency of the situation; they were already behind schedule on the release of the new graphics application that had been promised to several customers by a certain deadline. If Greg's team did not meet its deadline, the next software development team could not meet theirs, and the customers would receive the software behind schedule, which, based on contractual agreements, would result in a large financial penalty to his company. However, because of the way the company organization was set up, as project manager, Greg had no direct authority. In other words, his work was fully dependent on the team, but they didn't owe him consent or cooperation.

While observing the work practices of project managers, my colleagues and I noticed that everyone (not only Greg) employed a redundant communication strategy, although exactly how they did so differed based on whether the communicator had formal authority or not.

We found that people used two distinct forms of redundant communication to mobilize team members. Those who did have formal authority over their team members reactively followed an initial, often asynchronous communication attempt when their message that a threat existed failed to rapidly produce any change in employees' behavior. They would then launch a second communication, often synchronous, to make sure that their team understood that a threat really did exist. For example, a potential setback tended to result in short, "delayed" communications (such as email) intended to realign the work of employees to confront the new challenges. If employees seemingly failed to appreciate the nature of the setback and the need to make appropriate changes, the manager reacted by following up with redundant and usually "instantaneous" communication—for example, a conference call—to ensure that employees shared the managers' interpretation of events (and necessary workarounds).

Take Amanda, a leader at a large health care company. Early one morning, she learned that her company was changing its policy for insurance reimbursements for patient care providers. Charged with updating the reimbursement system for providers, Amanda realized that this latest policy change meant that her team would have to contact all of the providers whom they had recently transitioned to a new system. The providers would need to adopt yet another new version that reflected the policy change that she had learned about in that morning's meeting.

Because Amanda recognized that the new policy would be a setback, she sent Tim, who reported to her, an email sharing the new path that they needed to follow as soon as possible. She waited for some time to hear back from him and started to get nervous even though she realized that Tim was busy working on other projects. Did he not understand the urgency of the request? She followed her single delayed communication with a new instant communication. The instant follow-up signaled to Tim that this change in policy needed immediate attention over and above competing assignments. Later, in a chance hallway encounter, Tim told Amanda that he had not initially understood that this change in requirements was a threat to the project. It was only after the follow-up communication from Amanda that he came to see the event as urgently as she did.

Not everyone has Amanda's institutional authority. Like Greg, people who didn't have formal authority over their group engaged in proactive redundant communication to get others to focus on a collective effort. Project managers would begin with some form of synchronous medium, for example, the team meeting at which he offered people a chance to voice their concerns, that they would follow up with an asynchronous medium, such as email, that provided substance to the previously communicated message. Asynchronous mediums like the one that Greg labored over afford recipients the opportunity to process and digest a message that would otherwise be ephemeral or easy to dismiss. It also serves as a reminder of the importance of a task in a relatively nonintrusive manner since recipients can access the message at their leisure without having to respond in real time to a series of incessant demands.

Redundant communication is a reflection of how the judicious use of technology-mediated communication can help team members advance pressing work goals. Deliberate and strategic pairing of media will ensure that messages get through to recipients as befits their level of importance. While you'd think that individuals and managers work hard enough without having to communicate the same thing at least twice, in a world where people are suffering from information overload and messages can get lost or overlooked, redundant communication is an effective way to persuade people to move on things that matter to you rather than having to wait overly long before a response.

BRIDGING CULTURAL DIFFERENCES

Globally distributed teams often have members that come from vastly different cultural backgrounds. Does the same technology-mediated communication apply to groups that have diverse team members? One study found that cultural diversity hampered team communications and that using technologies appropriately could mitigate the negative impact of diversity on communication. Asynchronous modes like email lessened miscommunications that resulted from language differences. Synchronous communication enhanced trust and helped bolster a sense of team identity.

How do the contents of our communications interact with diversity and technology in distributed teams? In cross-cultural and cross-lingual settings, it may be tempting to send a simple email, "yes" or "no." To exchange information that requires subtlety or tact, another medium might be preferable. One study found that richer communication technologies allowed for the expressiveness necessary to exchange complex content. Leaner communication technologies allowed for the exchange of simple content while minimizing misinterpretations due to cultural differences. What might be regarded as common and considered appropriate in one culture might be rare and even anathema in another.

In global virtual teams, the cultural backgrounds of team members might also impact how they communicate using technology. Some cultures favor face-to-face interactions as a primary form of communication. Obviously, this is impossible for a globally distributed work team; therefore video-based communication as a top choice was important. If synchronous and visible media is not available, substituting with real-time audio conferences or phone calls is the best alternative. While emails can be used for exchanging trivial information, instant messaging platforms are a better option for individuals from cultures that commonly engage in "small talk" before diving into work. Unlike Western cultures where there is emphasis on holding real-time conversations for bad news, email turned out to be the better option to deliver bad news prior to a phone call so that the recipient had time to process asynchronously before the phone call.

Regardless of individual differences, we can't apply our beliefs and assumptions of what media is best across cultures. There are clear differences. My advice on how to get this right is always to ask communication partners for their preferences on the type of media to use for key work needs. Ultimately, the appropriate communication technology will be determined by the cultural and language backgrounds of our colleagues.

MAKE SOCIAL TOOLS WORK

Our modern lives are defined, in part, by the constant connectivity of social media personally and, increasingly, professionally as well. The fastest-growing company-based social media providers boast that they have tens of millions of daily users on their platforms. When they are implemented successfully, social tools allow otherwise distrustful workers to connect, share knowledge, collaborate, and innovate more effectively. Social tools also allow employees to learn about existing projects and initiatives that overlap with their own, as well as to coordinate efforts. This can reduce work duplication and free up resources to focus elsewhere where they are needed.

Consider the group of engineers at a high-tech multinational company whose conversations via social tools enabled an organic transfer of useful knowledge. When an engineer in the German office learned about a web analytics application that the more advanced Tokyo office had implemented locally, he contacted a Tokyo engineer for detailed information about the application and about required network support, adopted the application, and posted his satisfaction with it to the group. American and French engineers who read the post then expressed interest in the application for their local markets. Observing its success in Tokyo and Germany and its potential elsewhere, the group manager required that it be adopted across all markets. Similar company-wide spreads of knowledge in the marketing, sales, and legal groups also occurred as a consequence of spontaneous conversations over social media.

Employees who work in different locations around the world often have a hard time building relationships and forging a shared identity. Social tools can facilitate personal and professional connections, increasing trust and rapport across geographic and cultural borders. Many global employees report that internal social tools offer a window into broader organizational discourse that is otherwise unavailable to them. As an employee working for an e-commerce company explained, "I get a feel for what everyone's doing over there [at headquarters], the types of projects, and how they're doing. So I definitely feel more connected." Others at the company echoed his sentiment, saying, "I feel like I am part of the family," and, "We're the same company. We're the same people. We look different, we might sound different, but we're doing the same thing at the end of the day." In a virtual workplace where people rarely see each other, if ever, social tools can enable employees to feel a sense of belonging.

Bringing social tools into a company may look simple—and the technology aspect of it is. Slack or Microsoft Teams are examples of social tools that might be icons on your computer as they are for millions of people. Most social tools are cloud-based applications, so they require virtually no investment in infrastructure. Because most employees have had exposure to social media in one form or another, typically in their personal lives, learning how to use them at work tends to be easy. But despite this apparent simplicity, there are a few important things that people need to understand in order to reap their benefits.

In a longitudinal investigation of social users, my colleague Paul Leonardi and I compared social media implementation and user behavior at two companies. I like to call Paul my "academic sibling" because we were in the PhD program at Stanford University at the same time. Since then, we have coauthored a number of projects and have now collaborated for nearly twenty years on issues of work, technology, and organizations. In a financial services firm with over fifteen thousand employees, we looked at two departments over eighteen months. One department implemented social tools, and the other department did not. In the second company, a high-tech firm with over ten thousand employees in ten countries, we tracked the implementation and adoption of social tools across the company for more than twenty-four months.

What we learned was fascinating. Initially, participants would share non-work-related content in tandem with content that referred to "company-wide affairs." Coworkers became intrigued by the personal content shared by their colleagues, and this curiosity led to engagement with the site—posting and browsing both work-related and nonwork content. The visible intermingling of work and nonwork content, including public posts between coworkers, allowed individual employees to get a sense for whether they could trust one another even though they did not directly interact or share a meaningful affiliation. As discussed in chapter 2, trust allowed employees to determine if they could ask each other for help or share helpful knowledge regarding work-related matters.

In this way, mixing work-related and non-work-related content facilitated professional knowledge sharing by establishing trust. However, while this intermingling was initially beneficial for knowledge sharing within the organization, using the sites for nonwork matters ultimately led to anxiety and conflict: employees were concerned that managers would think they were socializing excessively, and interpersonal tensions were sometimes on display. These problems reduced people's engagement in non-work-related content, which also reduced the work-related knowledge sharing that had accompanied nonwork content. In sum, sharing nonwork content was paradoxical: it was the reason for both the rise and demise of social tool usage.

It's crucial for organizations to articulate why and how their employees should use the internal social tools, including allowing them to have nonwork exchanges. What do employees and the organization as a whole stand to gain? Unsurprisingly, leaders have to show the way. Setting guidelines for how to use social tools is not sufficient. Leaders have to be present on internal social tools, modeling desired behavior, for employees to do the same. For example, leaders can publicly engage an individual who posts a good idea, perhaps by asking a follow-up question. Leaders can also comment on non-work-related posts, such as wishing an employee happy birthday or "liking" a TV show. If, as happens often, leaders restrict their posts to mostly formal announcements about changes in policy or personnel, employees will get the idea that the tools are merely a way for management to broadcast information and will refrain from engagement and communication. The organization will fail to achieve its purpose for implementing social tools.

Context is everything when it comes to communication, virtual or not. It's not enough to subscribe to the latest social media platform or fanciest videoconferencing hardware. Sometimes it's better to wait on sending an email—or not send it at all—and other times it's best to press send ASAP. Employees and leaders need to become more strategic and mindful in deploying digital tools, and part of that process is learning to understand communication media in terms of its characteristics—lean versus rich, synchronous versus asynchronous—and apply that to what we know about relationships between the people with whom we work. As it turns out, a lot of our instincts about using these tools are wrong and produce a different effect than we intend. And the race to do more, more, more just because the technology allows is perhaps the most counterproductive. Finally, creating urgency and setting priorities is the job of the leader, not the technology. We may communicate virtually, but we are affected by the human aspects of social dynamics and social presence that we know from face-to-face interactions. Too many times, leaders lean on technologies to do the work of setting priorities and getting a feel for what's going on with their teams. They are not a substitute for the work of leading.

Success from Anywhere: Using the Right Digital Tools

- **Mix it up.** Tech exhaustion occurs when we let digital tools structure communication activities such as scheduling too many videoconference calls back-to-back, rather than structuring activities around our own needs. Using a mix of available media—synchronous and asynchronous—to match our goals lessens tech exhaustion.

- **Understand context.** The mutual knowledge problem is a natural yet detrimental aspect of working remotely and relying on digital tools to communicate. As a result, remote workers will miss situational contexts and may be on unequal footing when it comes to shared information or assumptions. This reduced common ground can create misunderstandings and obstacles to productive collaboration.

- **Be present.** The social presence problem stems from the extent to which certain digital tools enable us to deliver social cues that promote intimacy—the feelings of interpersonal closeness—and immediacy—the psychological distance or feeling of mental or emotional connection between speakers.

- **Remember that less can be more . . . and vice versa.** The richer media will be more effective in situations with higher ambiguity, higher equivocality, and less clarity, while leaner media will be more effective in situations that are straightforward.

- **Repeat yourself . . . strategically.** Depending on their degree of authority, people can sequence the media they use to communicate redundantly—from synchronous to asynchronous or vice versa—thereby conveying the importance of a message or making a call to immediate action.

- **Don't forget to ask.** Cross-national teams need to take into account cultural and language differences. Preferences for synchronous or asynchronous communication differ by culture and common language competence.

- **Close social distances.** Social tools help far-flung coworkers connect, share knowledge, collaborate, and innovate more effectively, as well as reduce work duplication and free up resources to focus elsewhere where needed. Nonwork communication on social tools lubricates work communication, and both leaders and employees should engage in social chatter on company-wide social tools.

CHAPTER 5

How Can My Agile Team Operate Remotely?

In the popular sitcom TV series *Silicon Valley*, about a group of six software developers working on what they hope will be the next big product in Silicon Valley, California, the members of what's essentially an agile team live together in one house. Their impromptu conversations to hash out issues as they arise—be they tech related, logistical, or interpersonal—are as apt to take place in the kitchen, driveway, hallway, or yard as they are at scheduled meetings in their living room office. In other words, the team members are constantly together in the same physical space. Collocation, according to the show's set, determines the team's close, innovative, and dynamic collaboration and enables their passion and drive.

The comedy series, created by Mike Judge, John Altschuler, and Dave Krinsky, draws heavily on the real-life software culture in which agile teams originated. Software developers, and the nature of the computer code they write, need working methods that support collaborative, team-driven work to bring new software products quickly to market. That's why, in the late nineties, the steep growth in systems programming spawned an urgent need to update the traditional "waterfall" method of product development in which highly structured plans for task completion are drafted before development takes place, and then passed along sequentially to specific departments at each stage. Among other critiques, this highly structured and time-consuming approach meant that products could be outdated before they were ready to be delivered to customers.

In 2001, seventeen leading software developers met at a lodge in Snowbird, Utah, to talk, ski, eat, and come up with a new method for software development wherein teams could get products in the hands of customers in the earliest stages of the process. The sequentially driven waterfall approaches that the seventeen software developers sought to replace had been around since World War I, when consultant Henry Gantt created an organizing mechanism for U.S. military strategy. In the words of Jeff Sutherland, one of the authors of the Agile Manifesto, "We gave up on trench warfare, but somehow the ideas that organized it are still popular."

The meeting, itself an example of the power of in-person collaboration, resulted in the "Manifesto for Agile Software Development," which described the new adaptive and iterative approach in succinct terms:

We are uncovering better ways of developing software by doing it and helping others do it.

Through this work we have come to value:

Individuals and interactions over processes and tools
Working software over comprehensive documentation
Customer collaboration over contract negotiation
Responding to change over following a plan

That is, while there is value in the items on the right, we value the items on the left more.

© 2001, Agile Manifesto Authors
this declaration may be freely copied in any form, but only in its entirety
through this notice.

Since 2001, agile teams have proliferated, and have spread well beyond the software industry in Silicon Valley; the term has acquired a bit of buzz and mystique. To set the stage, this chapter first explains how agile teams are designed and operate, and then asks how the agile philosophy—which prioritizes close and frequent face-to-face interactions throughout the working day—can go remote, if at all. Although agile and remote might seem a contradiction—and in some circles, even a blasphemy—in fact, I have found that agile teams and remote work are surprisingly well aligned. To show you how this can be done, we'll look first at London-based Unilever, one of the largest global companies in the world, and discuss how its digital transformation strategy operates with remote agile teams at scale. Finally, you will learn how AppFolio, a midsize software company located in California, managed their sudden transition from collocated agile to remote work.

DESIGN OF AGILE TEAMS

The design of agile teams is built on the core principle that competitive advantage lies in the flexibility to find the most effective configuration of resources and capabilities. Teams are small in size to enable fast decision-making and high productivity; too many people and complicated communication streams can easily overwhelm the team and slow things down. Most agile experts recommend five to nine people on a team as optimal. Roles are fluid, and team composition is cross-functional so that team members can take on any of the tasks involved. Decision-making tends to be shared, and no one person is "in charge." Teams self-organize around challenging and compelling goals, creating an urgency to "get things done," which increases energy and motivation and keeps team members engaged. Team members who assume ownership over their tasks and are trusted with decisions about how to execute from inception to completion are comfortable with a high degree of autonomy and accountability.

Open, direct, and frequent communication is central to the agile method, enabling individuals to quickly raise issues to the larger team and work with managers to find solutions. Because teams thrive when they can quickly see the results of their work, agile teams use quick experiments to capture feedback from internal or external customers and make decisions accordingly. As customer needs continue to evolve once a product or project is kicked off, rapid prototyping and continuous collaboration with customers help ensure that end products deliver real value. With such an iterative approach, detailed up-front plans or lengthy documentation after the fact is not useful. Instead, agile teams agree on a vision and direction to start a project and expect to adapt the tasks as they go.

Since their founding, perhaps the most definitive characteristic of agile teams has been the insistence that team members meet with high frequency. Agile teams became known for daily meetings, often held at the same time, where each person shares a brief progress report with the rest of the team. While the frequency of the meetings can vary in different environments, the meetings are intended to be regular and brief—no more than fifteen minutes. Everyone on the team is expected to participate. The tone of these meetings is positive, with the team identifying what's working and what's not, and determining how to remove any obstacles to progress going forward. High trust, candid conversation, and accountability are critical for real learning and innovation to happen.

Agile is premised on collocation. It's difficult to overstate this assumption. The Manifesto states explicitly, "The most efficient and effective method of conveying information to and within a development team is face-to-face conversation." The belief is that face-to-face communication makes teams more agile because it eliminates the confusion and overhead often caused by excessive documentation. Team members' frequent in-person contact throughout the day allows for quick check-ins, corrections, and bonding. Drafting a document, another waterfall practice, is discouraged in agile because it takes extended time, is often superfluous, and its contents could be misinterpreted by readers without the author present to clarify. Face-to-face conversation has been seen as the gold standard that can best resolve any misunderstandings on the spot in real time through a collaborative back-and-forth.

These key features of the agile process seem to make it incompatible with distributed teams or people working remotely. However, as you will see later in the chapter, agile processes have been scaled to globally distributed teams with great success, and have also thrived in a remote format—even for collocated agile teams suddenly ordered to stay at home as the coronavirus raged. Remote teams and managers can take heart—if a methodology that's so intensely dependent on daily, collocated meetings can be transformed to remote success, then anything can.

BEYOND SOFTWARE

The Agile Manifesto originated in software development, but since then its methods have become appealing in a world where product cycles are shorter and shorter, and information becomes more and more plentiful. Although born in an environment shaped in large part by the rapid development of digital technology, agile development is not about specific tools or disciplines. Agile is about *how* teams work in an environment. The tools, frameworks, and processes that agile offers are applicable beyond software development and technical teams. Let's briefly consider, for example, the ways in which a toy manufacturer, an R&D team, a radio show, and two banks adopted various aspects of agile methodology to positive outcomes.

Toy manufacturer LEGO adopted agile because they wanted to increase visibility into product development processes across the company. First, they formed product teams to work as self-organizing and autonomous scrum teams that learned via iteration. Then a group of teams began meeting every eight weeks to showcase their work, work out dependencies, estimate risks, and plan for the next release period. Finally, LEGO created an agile layer for top management and stakeholders to ensure the work was connecting to long-term business objectives. Developers managed their own work, and were able to give more accurate estimates for product delivery. Agile methodology led to more predictable and positive outcomes for the company.

Members of 3M's research and development team must constantly imagine, create, prototype, and refine new and innovative approaches to product development for the multinational conglomerate operating across numerous industries. Each of these steps is inherently time consuming. When 3M adopted the scrum framework for the new product development, it adjusted the meeting frequency, the documentation process, and other agile methods to better suit researchers' needs. These changes enabled researchers to balance the strictly mandated project deadlines with the flexibility that innovation requires. Projects were broken down into small steps with flexible expectations. As a result, the team accomplished more work for 3M with less stress and improved efficiency.

Regardless of the industry, keeping the end customer's needs at the center of the process at every stage is essential to the agile approach. The goal is to ensure that the delivered product or service provides value. Agile teams write their tasks as "stories" in the language an end customer would use, answering: *Who is the task being done for? What do we want done? Why does the customer want this?* Frequent sharing and feedback loops that inform the next set of "stories" help teams stay focused on building products according to the customer's needs and expectations and are intended to replace guesswork or theoretical forecasting for product sales.

National Public Radio used this particular aspect of agile to create new shows with less expense and risk. In the past, the network had created new programming that involved big, expensive launches with no underlying data to ensure the program's success. Agile allowed them to distribute small pilot shows to its radio stations instead. Teams would gather feedback from both local program directors and listeners to quickly determine a particular show's success or failure. By developing the pilots that resonated with its listeners and cutting the ones that did not land, NPR was able to achieve significant cost savings while growing its audience.

Santander's marketing team began experimenting with ways to create value out of its banking data. Instead of long marketing cycles led by agencies, the bank launched small, low-risk marketing campaigns released in two-week sprints. The company learned immediately which campaigns were successful. This new intelligence has helped the bank reach customers at specific times with targeted content. In one recent trial, loyalty increased by 12 percent, and account satisfaction increased by 10 percent. The bank's Net Promoter Score has reached its highest point in seventeen years.

ING in the Netherlands adopted agile methodology to shorten time to market, improve customer experience, and improve operations and digital banking capabilities. The bank radically restructured the Dutch headquarters, cutting 25 percent of the workforce in the process. The agile approach relied on small, multidisciplinary, autonomous "squads" that were responsible for customer service from cradle to grave. ING's implementation of agile led to faster service, broke down organizational silos, significantly reduced the number of handovers, and increased employee satisfaction.

These adaptions and adoptions are examples of the agile team methodology's spread from the relatively insular world of software developers to contemporary business management across a range of industries. In each case, the increased performance outcomes depended on the agile premise of small, autonomous, self-organizing, and cross-functional teams whose members frequently collaborated in person with one another. But the world has changed since 2001, when the Agile Manifesto was written. In-person collaboration is not always possible or even desirable. The fact that customers are global has already pushed many agile teams to collaborate virtually, across national boundaries.

Today the challenge is whether agile teams can stay agile without face-to-face interaction. In other words, how can teams reconcile the methods and needs of agile with the methods and needs of remote teams, with all their attendant issues of trust and communication? The rest of this chapter will demonstrate how this can be done.

UNILEVER'S DIGITAL TRANSFORMATION AGILE TEAMS

The organization is less like a giant warship and more like a flotilla of tiny speedboats . . . an organic living network of high-performance teams.
—STEPHEN DENNING, AGE OF AGILE

Since 2017, Rahul Welde has been instrumental in implementing more than three hundred agile teams that operate remotely across so many time zones that its remote town hall meetings are known to begin with "Good Morning, Good Afternoon, and Good Evening." Unilever is a London-based multinational consumer goods company, and what began as a pioneering agile initiative in consumer marketing, where teams reorganized themselves to be autonomous and leverage "empowerment, collaboration, and agility," soon evolved into a format for the broader organization. Welde, Unilever's executive vice president in digital transformation and a twenty-nine-year veteran of the organization, was convinced that these seemingly opposed approaches to teamwork—agile and remote—could be combined and were in fact necessary as the company underwent a digital transformation strategy. Unilever's approach is testimony to agile teams' ability to be remote, global, and operate at scale. Moreover, Welde found that the marriage of agile and remote is ultimately a marriage of digital and global.

The Global-Local Dynamic

With more than 400 brands across 190 countries, Unilever has relied on remote work for decades, and built its global organization around a distributed team structure to reach a wide range of multiple markets. For a multinational company that makes and sells household staples—from brands like Dove soap to Magnum ice cream bars—success is defined by a delicate balancing act between the specificities of local markets and the broad scale of global operations.

As Welde puts it, "Our consumers are very local in nature . . . while we have an overall framework, we still have to bring it to life for a given brand or to a given geography. As an example, what we do in China would be very distinct from what we do in the U.S., and would be very distinct from what we do in the U.K. Or for that matter, what we do on Dove would be very different from what we do on Lipton tea or Magnum ice cream."

To meet the particular conditions under which they must operate, Welde told me that he sought to build new ways of working "glocally" by using digital technologies such as cloud computing and big data to develop relationships with local markets. Having had the privilege of spending some time hearing Welde speak about his vision and later sitting down with him on several occasions, it is easy to see that he is one of those special leaders with the capacity to see near and far, think fast and slow, and someone who possesses the courage to go agile full-on for the benefit of customers across the globe.

Consumption of goods is not a global act, however. The actual point of sale occurs as a local experience and is dependent on what happens on the last mile before the product reaches its final destination in a town, a shop, and ultimately, its placement on a shelf. Agile, results-oriented teams that worked iteratively, autonomously, and remotely could focus on the unique demands of a particular last mile while simultaneously guiding their work with the company's digital capacities across multiple countries. Welde realized that this digitally driven interplay between local magic and global scale was the key element that remote agile teamwork could offer.

For the tech-driven startups such as Cisco and Oracle that had relied on agile methods to develop, global was the natural growth outward. For Unilever, a ninety-year-old company whose "bread and butter" was not an application or software, but quite literally bread and butter, the trajectory was reversed: Unilever needed to figure out how to take a gargantuan and sprawling global business into the digital age. Unsurprisingly, the first of three vectors of this monumental task was technology, or enabling tools. The second was process, a company key strength that they learned to rewire to adapt to new technology and tools. Last and most important of the vectors was people. People consume the many products that Unilever provides—whether it is experiencing a soap's scent, tasting ice cream, or drinking tea—and people were behind the work of taking each of these goods from an idea to reality. Agile methodology—with its emphasis on digital technology, iterative processes, and close collaboration—informed all three vectors.

Unilever's transformation to digital technology was what propelled its adoption of agile teams. Most astonishingly, the synergy between both these innovations enabled a sprawling, legacy multinational to bridge the global-local divide.

AppFolio: Born Agile

Unlike Unilever, AppFolio was *born* digital. The ethos was written into its founding mission: develop software to help certain vertical industries—real estate, for example—transition into the digital age. The company's first product was a software solution for property managers. In most respects, it followed the orthodoxy of the Agile Manifesto from its founding. Eric Hawkins, director of engineering at AppFolio, attributes the company's success to its agile teamwork structure. One of the company's values says, "Great people make a great company." Hawkins believes that small, focused teams keep them agile.

I first became interested in AppFolio through my colleague Paul Leonardi, who is currently a professor of technology management and engineering at the University of California, Santa Barbara, and knew many of the AppFolio teams, including one of the founders—former UC Santa Barbara computer science professor Klaus Schauser, who had formed AppFolio in 2006 along with tech startup veteran Jon Walker.

Schauser and Walker's awareness that businesses need innovative software to evolve into the digital age had inspired them to start a company, and the agile philosophy of constant iteration aligned with their mission. They structured their small company around agile teams for all project-based work, from software development to marketing. For fourteen years, their company could have been considered a model for how to how to effectively design and use agile teams.

Each agile team was composed of a product manager, a designer, a quality assurance engineer, and a few full-stack software engineers who could fill any role in the development process and switch between teams depending on the needs of specific projects. Product managers spanned about two teams, and were overseen by a product leader in each team. Each team had the autonomy to select its own projects, and did so through an immersive process of face-to-face brainstorming. Instead of being assigned narrow tasks to complete, teams were introduced to a vague overarching problem that gave them the freedom to iterate, improvise on the fly, and build off one another toward their own solution. As a result, AppFolio's agile teams were motivated, fast, and stayed in close touch with customers. This autonomy was also a key factor in the company's talent acquisition: because top talent always had the option of joining big players like Google, App-Folio's appeal was the opportunity to work on challenging technical problems that small agile teams chose themselves.

True to the Agile Manifesto, each morning teams held a "stand-up" meeting—a face-to-face huddle in which the team leaders checked in with team members to talk about how the work was progressing. As Hawkins put it, "We believe face-to-face conversation is the highest bandwidth form of communication. It allows for the quick decision making that is necessary in agile development." Hawkins, who oversaw six agile teams, also held one-on-one meetings with each of his twenty-five direct reports every week. To encourage a collaborative exchange, he often held these meetings outdoors on a casual side-by-side walk in the California sunlight instead of sitting across from one another at a desk.

In the spirit of agile, Hawkins often took full advantage of his engineers' full-stack capabilities by regularly moving them between teams to make sure every project had the best lineup for its specific goals. As team members moved around, they got to know one another and expanded their familiarity with other AppFolio employees. The result was a network of interconnected teams all working together at the Santa Barbara headquarters—a free-flowing, open-concept space that encouraged cross-pollination of ideas. Clayton Taylor, a user experience designer at AppFolio, saw this as an essential element of the company's success with the agile approach: "Being co-located gives everyone at least a baseline knowledge of all the projects going on. There's a natural awareness by proximity." For example, if Tyler's team was looking for outside advice, he could walk a few feet and ask any number of colleagues he had gotten to know personally on previous agile teams. Conversations at the office would often continue after work over drinks.

Hawkins characterized the collaborative atmosphere at the App-Folio office as "interrupt-driven," where his role as a leader was to be present but not overbearing—available whenever his team members needed guidance, but otherwise hands-off. Hawkins valued the opportunity to see his teammates at the office. He espoused an open-door policy where anyone could drop by his desk to pose a question or ask for help. He viewed his job as an iterative process of responding to his team's needs as they came up in real time.

AppFolio Goes Remote

Like so many companies across the world, AppFolio's entire existence abruptly and dramatically changed in 2020. When COVID-19 put the United States on lockdown overnight, AppFolio's sudden shift to remote work posed a direct challenge to the company's agile in-person practices of teamwork. Although Hawkins and his team initially faced the change with optimism, one week in a remote format had taken an immediate toll. As Hawkins put it, "We hit the ground running. There was a lot of initial enthusiasm and energy. People were like, 'Right, we're doing this. We're all in it together. We're working from home and we're just going to keep going.' And by the end of that week, I was just shattered. I was so exhausted. Early the next week, I asked others how they were feeling, and they were like, 'Oh boy, I don't know how many more of these back-to-back videoconferences I can do.'"

Hawkins realized right away that their accustomed way of collaborating did not translate into a virtual context. All the ingredients that went into the collaborative "interrupt-driven" culture of AppFolio's agile teamwork, as Hawkins called it, suddenly disappeared. No longer could they spontaneously huddle up to have a quick conversation when a problem arose. The office had been built to support a process that involved, according to Hawkins, "zero friction." Now everyone was huddled separately in their individual homes, working under the new remote conditions. Videoconferencing had replaced a walk in the sunlight.

AppFolio's agile teams thrived on the natural rhythms of informal interactions when working in the same office. As Tyler pointed out, collocation was crucial to the teams' freedom to improvise and innovate through organic exchanges. A casual conversation about a Netflix show might lead to an intense brainstorming session about a new project. But in the virtual mode, these natural rhythms were inaccessible. People's interactions were now confined to a combination of text, audio, and video at discrete times of the day. Casual conversation and nonverbal cues such as hand gestures and facial expressions—key elements of face-to-face interaction in the workplace—were lost in translation.

Hawkins noticed that the absence of these nonverbal cues made it more difficult for team members to know when to speak up and when to listen. As a result, teammates often unintentionally spoke over one another and virtual meetings began to feel crowded. On the flip side, one-on-one meetings in a virtual setting could feel stilted and unnatural—the polar opposite of his informal outdoor walks with team members when they were collocated at the Santa Barbara headquarters.

Those first two weeks working away from the office were tough. Eventually, though, AppFolio found agile ways to work remotely. In some cases, as the team realized, working from home can have distinct advantages. The transition they made demonstrates why, given the choice, many people may opt for a hybrid version of working remotely from home and commuting into the office.

As AppFolio's transition into remote work illustrates, agile methods and remote work are *not* incompatible despite what the original doctrine may assert. In many ways, agile teams can maintain the spirit of the Manifesto while adapting some of its practices. Hawkins admitted that only 10 to 20 percent of what his teams must accomplish in a given week are truly collaborative or creative tasks. The rest is really individual, focused work. To his surprise, Hawkins found that his team members were much more efficient and productive accomplishing the focused work *without* the distraction of being close to one another at the office. He's come to think that this benefit may outweigh the cost of losing the collaborative activities his teams had long practiced.

EASING THE TRANSITION TO REMOTE WORK

In general, I've found that teams that had previously established individual norms for communication or other ways of working together were well-primed for the transition to remote work. For example, the members of one agile team were accustomed to vocalizing individual preferences about what to do when wearing headphones. One team member had instructed the others to "just tap me on the shoulder" to get his attention. Another had stated a preference that fellow team members "shoot me a message on Slack and see if I can pause what I'm doing" before interrupting with a question. This level of comfort with one another around individual needs translated well to remote work, enabling team members to be clear about what types of scheduling or communication worked best for them working from home.

Teams that may have incorporated digital platforms into meetings where some members are collocated while others are virtual are also primed for extensive remote work. For example, a multinational company headquartered in the United States had long relied on scheduling specific conference rooms in their building for the larger sprint team. Although the company was ostensibly collocated, approximately 30 percent of the team participated remotely—because they worked in offices in another state, another country, or simply because they needed to be home that day to, for example, wait for the plumber to arrive—by dialing into a specific room number. Technical problems arose for the remote team members trying to sign in if the room had to be changed at the last minute due to scheduling constraints for the conference rooms, which were typically in high demand.

However, shortly before the pandemic shutdown, the agile team had switched to using an internal social media platform that integrates with email clients and people's mobile phones. The platform allowed everyone to dial in remotely to a specific number rather than to a specific physical location, eliminating previous technical snafus. Once the team moved to full-time remote work, the norms for scheduling meetings were already established, making the transition that much easier. Team members had already made the mental shift from scheduling a meeting in a specific place—the conference room—to a specific time accessible by a common dial-in number.

BEST PRACTICES FOR REMOTE AGILE TEAMS

Among the remote agile teams I studied—both established and newly formed—I found five common practices that allow people to generate and maintain productive collaborative energy in a remote format. In each of these practices, the inherent assets of remote work—such as efficiency and speed—were not only compatible with the agile method, but directly aligned. Remote agile teams are not second-rate to collocated agile teams; with some adjustments and in some cases, agile principles can be better served by teams who do *not* work face-to-face out of a common physical office.

Prepare Alone, End in Sync

Adapting agile methods to a remote context calls for a transition from constant collaboration into practices that combine self-directed solo tasks on one's own schedule with real-time collaboration efforts. That is, remote work requires team members to each work asynchronously in order to lubricate the agile process of spontaneous face-to-face collaboration. Spending individual time in pre-work or pre-thinking on matters that previously could have been resolved in real time becomes paramount. Sending out a simple agenda prior to electronic meetings or asking team members to reflect on key items before convening helps maintain the short and efficient meeting processes that agile approaches require.

Virtual meeting platforms fail to provide the natural conditions for real-time brainstorming. As a result, asking team members to jot down thoughts on a shared platform prior to group brainstorming is an important shift in remote agile collaboration. As an initial step for proposing ideas, a team can use any asynchronous form of communication it may be accustomed to. For example, before a real-time virtual meeting, team members can compose their ideas in emails, internal social media, or shareable documents that the rest of the team can review and comment on. When the team convenes, members can immediately start appraising certain ideas or home in on challenges that they need to resolve rather than spend valuable time hashing them out in the first place.

Brainstorm in Shared Documents

Interestingly, in conversations with agile teams that have gone from in-person to remote work, members express that virtual arrangements have brought the team *closer* to the agile ideal than collocation did. Using asynchronous collaboration tools, such as Google Docs, allows the team to constantly iterate without the "guardrails" or boundaries of a conventional collocated workday. Team members can make comments or suggestions on a shared artifact whenever a thought comes to them—on their own time—instead of waiting for the appropriate moment to broach the subject with colleagues in the office during set meetings, or when a colleague doesn't appear to be busy. For this reason, remote agile practices that call for a dedicated focus on interacting with actual artifacts undergoing constant iteration can be more aligned with the agile premise than the collocated practice of informal huddles on a whiteboard, which can't be saved for further discussion beyond a photograph.

For managers, this method is particularly useful to socialize an idea and get the team to make a decision quickly. If you have a collaborative idea that you want to buy in on, it's helpful to write it up on a short informal document, share the document with the team, and let people comment on it asynchronously. In other words, let the idea converge naturally as people engage with it on their own time. After everyone has had a chance to comment and offer input, managers can convene the team for a virtual meeting to discuss any lingering concerns or final comments. Because everyone has had a chance to communicate their ideas in written formats that are saved for future reference, coming to a decision is often much easier than hashing everything out in a collocated office.

Streamline the Huddle

The fulcrum of the collocated agile team is the daily stand-up meeting. Collocated teams' custom is to have someone present their work in the stand-up, and team members respond with input whenever an insight comes to them. People chime in, ad hoc style, on each piece of work. When everyone is sitting in a room together, its effectiveness depends in part on people's ability to read social cues and see that somebody else is about to speak. Clearly, that no longer works during virtual meetings.

Remote work requires new customs for conducting daily stand-up meetings. A little more orchestration is needed. One option is to give each person a dedicated time to speak without interruption before handing the virtual baton to the next person. This approach eliminates the problem of people unintentionally talking over one another or waiting to read a virtual "room" that is no longer readable.

One challenge that larger agile teams of up to nine or ten people encounter when working virtually is the ability to provide easy input using mediating technology. When meeting remotely, it's challenging to accommodate as many as ten voices interrupting and speaking simultaneously. Reducing the number of people involved in a virtual huddle in the early stages of a project can help focus communication. Holding meetings with a small, cross-functional group to include, for example, one engineer, one project manager, and one designer can accelerate decision-making as well. Once the smaller group has reached some preliminaries, more people can be brought in to contribute more opinions and higher energy.

Virtual meetings can be more efficient than face-to-face meetings. Although strict agile rules call for a daily meeting to last fifteen minutes, in reality teams can find it difficult to keep to that allotment. The "daily stand-up" is designed to give each team member two to three minutes for a progress report; so if a team has six members, it would optimally take no more than twelve to eighteen minutes to get through the meeting. But time is also spent waiting for the room to clear from a previous meeting, plugging computers into and out of power outlets, chatting after the meeting. When all is said and done, the meeting can last closer to thirty minutes rather than fifteen.

Virtual meetings, however, circumvent most of those complications. If people "arrive" early for a virtual meeting before the team is fully assembled, they can catch up on a short task such as email. Transitioning back to solo work can also be easier after a virtual meeting if one simply logs out of an application rather than having to walk out of a meeting room and back to a workstation.

Virtual meetings benefit from two types of digital tools that are less feasible for face-to-face meetings. The first is virtual whiteboards, which are easier to see on a screen than whiteboards in the physical office, where the view can be blocked if sitting at awkward angle. The second is screen sharing, which team members can toggle for a full view of specific people's work—much more efficient than looking over someone's shoulder at their computer when in the office.

Set Digital Norms

Remote teams establish norms that identify which digital communication platforms are most appropriate for certain forms of correspondence. For example, email might be deemed most suitable for formal but non-urgent requests, while instant mobile messaging applications might be more appropriate for a less formal but more urgent request.

Phone calls can be used for quick check-ins. Even if office environments are set up to support team members' accessibility for quick one-on-one chats or huddles, not every collocated agile team works in such congenial architectural settings. Especially if people are shifted from team to team as projects demand, they may be located down the hall from one another. Decades ago, before mobile phones came on the scene, employees used a corporate desk phone to check in with a team member located two floors above or three floors down. Digital technology ushered in chat rooms and instant messaging for communicating about informal matters that did not necessitate a walk down the hall and a knock on a team member's cubicle for an informal check-in.

Working remotely, using one's personal mobile phone to call a team member becomes a replacement for the quick cubicle check-in, with the slower and more laborious texting again relegated to matters that require little to no back-and-forth discussion. Especially if working from home involves young children, and people are sitting in front of their computer screens wearing noise-cancelling headphones, it's often easiest to make a quick, spontaneous call.

Because virtual communication can occur at any hour of the day or night, whereas face-to-face interaction at an office is bounded by more standardized work hours, managers need to establish guidelines on when to communicate—and more important, when *not* to communicate—in order to preserve the boundaries between work and nonwork responsibilities.

Solicit Anonymous Feedback

Agile team collaboration is predicated on the practice of honesty, trust, and candid communication. Team members talk with one another rather than directing comments upstream to a supervisory manager. Agile has retrospective reviews built into the end of each work segment—or "sprints"—during which team members can post anonymous Post-it notes on a specified office wall stating what they liked about the experience, what they didn't like, ideas sparked, and things to celebrate.

However, honesty, trust, and candid communication are not always easy to maintain for an intimate, collocated team, and even more difficult for agile times that are remote, large, numerous, or all three. Yet ongoing feedback and candid communication about team process and dynamics is still crucial. On remote teams, leaders can use interactive tools to gather real-time data about people's experience. For example, during a virtual meeting, colleagues can submit anonymous questions, thoughts, or concerns about the discussion topic. Simultaneously, the team leader can run an anonymous poll to capture colleagues' opinions.

Anonymous features of digital tools allow for comments to be especially candid, which in turn can help teams learn from mistakes and improve without fear of repercussions. Team members who are unhappy about a particular issue can voice concerns through various polls. Word clouds made up of such anonymous responses can be aired in the middle of a meeting to stimulate conversation or to solicit instant feedback. These digital tools afford opportunities that in-person interactions can inhibit. People are often hesitant to provide unfiltered thoughts and comments in a group. Across multiple agile teams, real-time feedback can inform productivity at the team level and at the broader department level within the organization, offering a wide scale of analysis.

Agile principles can be served by teams who do *not* work face-to-face out of a common physical office. Traditionally, a core requirement to agile work was small, collocated teams that were thought to best enable brief, daily meetings; everyone reported on progress to date, discussed issues as they arose, and collaborated on next steps. Multinational firms, among the first to see how agile teams could work remotely, have successfully scaled agile methods and philosophy among distributed teams. With some adjustments and in some cases, remote agile teams are not second-rate to collocated agile teams.

Success from Anywhere: Agile Remote Teams

- **Prepare for virtual meetings asynchronously.** Brainstorming in emails or group documents prior to the meeting enables spontaneous collaboration. Because everyone has had a chance to communicate their ideas in written formats, which are saved for future reference, coming to a decision is often much easier than hashing everything out in a collocated office.

- **Deliberately orchestrate daily or frequent meetings.** Give each person a dedicated time to speak without interruption before handing the virtual baton to the next person. Preliminary meetings can be held with a small, cross-functional group before bringing in the larger team for more discussion.

- **Take advantages that virtual meetings uniquely afford.** Individuals have control over their asynchronous work time that helps the team. Make use of virtual whiteboards and toggling screen sharing to increase efficiency.

- **Frequent launches and relaunches are crucial.** (As described in chapter 1) Because remote team members rely on digital communication—phone, email, text, and videoconferencing—individuals need to make the effort to stay in contact, and the group needs to set norms about what kind of communication to use and when.

- **Collaborate using digital tools for continuity.** Unlike huddles in front of a whiteboard or water cooler, team members who use digital tools to communicate can capture the output of their work. Rather than fizzling into thin air, artifacts emerge that can be modified, improved, or revisited for subsequent group work.

CHAPTER 6

How Can My Global Team Succeed Across Differences?

If you were raised in a North American culture you were probably taught that making eye contact when talking to another person projects confidence and honesty. If you were raised in other parts of the world you may find direct eye contact rude or threatening, especially if you don't know the other person well. When team members raised from these two different cultures work together, the North American may unwittingly make his colleague feel uncomfortable. The team member who is not used to direct eye contact may unwittingly project an air of disengagement from the project when that is not at all the case. This is just one small example of the cultural differences that affect global teams who work across borders. How we greet one another, close a deal, make decisions, speak to figures of authority, and more are all behaviors that subscribe to cultural norms that differ depending on where you are located in the world.

Cultural differences are inherent in remote, global teams. As discussed in chapter 2, the interplay between how we see ourselves and how others see us is a dynamic process that influences our behaviors and emotions. We often find it easiest to align our self-perceptions—eye contact projects confidence—and others' perceptions of ourselves—eye contact is threatening—by surrounding ourselves with other people who think similarly, but that becomes impossible when working on a global team comprising people from different cultural backgrounds. We are constantly, through our interactions, giving clues that demonstrate how we feel we should be perceived. Someone who expects to be perceived by her team members as a leader may elicit support from long-term colleagues, make frequent reference to her own skills and experience, and take informal control over team processes. In a global environment that by its very nature includes cultural differences, finding a way to negotiate and balance our own and others' perceptions of who we are can be a significant challenge. If unaddressed, cultural differences can erode team morale, shatter trust, create discord, and lower performance.

Organizations often adopt inspirational mottoes or set aside an annual event to celebrate diversity as go-to remedies to cultural trouble. These gestures are important, but they don't solve the day-to-day problems that arise for globally distributed teams that are likely to have members who come from many nationalities. Building positive, productive common ground with mutual trust and understanding—as opposed to merely abstaining from insensitive remarks or actions— must be undertaken on an ongoing basis. You can't simply memorize a list of "dos and don'ts" for each respective culture—both because it's a limited way to build understanding and because you quickly enter the realm of stereotype. Not everyone from one culture will necessarily hold one set of values or behave in the same way. Instead, teams must do the deeper work to develop an understanding both of how others see the world and how they perceive your own behaviors.

In this chapter, you will learn how to approach the deeper work that allows you and everyone else on your team to work across cultural differences. First, in order to understand the extent of the problems that language and cultural differences present for globally distributed teams, we'll take a comprehensive look into the predicament faced by

Tariq Khan, a new manager hired by a multinational petrochemical company with an employee pool so diverse that eighteen languages were spoken. Their failures across differences were chronic and sticky. Next, you will learn about the history and consequences of what sociologists call psychological distance, a root cause that global teams must address, and then how Khan managed to turn his contentious and low-performing global team around. The last part of the chapter provides an extensive discussion of mitigating actions and approaches that I use when working with leaders of global teams.

27 COUNTRIES, 18 LANGUAGES, AND 1 FAILING TEAM

Late in the evening at Tek's Dubai office, Tariq Khan sat across the conference table from three senior executives who had spent the last sixteen hours arguing about why their global sales and marketing team had fallen apart so disastrously. The petrochemical company had recently offered Khan the high-profile job of leading a large globally distributed group, which consisted of 68 members who originated from 27 countries, spoke 18 languages (including many other dialects), and whose ages spanned from 22 to 61 years old. The job offer spoke to Khan's high leadership potential and precocious success in his previous roles at Tek.

However, the stakes were high. The odds were stacked against Khan. In two short years, the team's operating margins had dropped from 61 percent to 48 percent, net profit margins from $46 million to $35 million, and market share from 27 percent to 22 percent. Employee satisfaction had also fallen from 68 percent to 36 percent. The departing manager had resigned in disgrace. His parting words to Khan were stark: "Listen, Tariq, I am going to be completely honest with you—

the situation is simply out of control. This job has ruined my reputation here, I have no choice but to leave. If I were you, I would think twice before taking this on."

The previous manager's ominous words echoed in Khan's head throughout his discussion with the senior executives that evening in Dubai. He had scheduled the day-long meeting with the executives—Sunil, Lars, and Ramazan—thinking he would get answers, but their competing theories only raised more questions.

The discord had been palpable from the minute Khan stepped into the conference room that morning. Each executive had his own explanation for the team's precipitous decline. Sunil, an Indian national working in Lebanon, blamed the group's recent financial failures on the market, claiming that price increases in base oil were putting pressure on the team's margins. Lars, an expat from Sweden, vehemently disagreed. In an accusatory tone, he blamed their troubles on bad branding that confused customers and failed shipments that chipped away at the relationships with Iranian and Yemeni partners.

Sunil gave no indication he'd even heard what Lars said; instead, he changed the subject and began critiquing the team's compensation structure. He pointed out that the variable portion of salaries was based on volume and revenue instead of earnings or margins; so when prices went up, salespeople were able to make revenue targets by selling less volume while the cost of goods sold was still increasing and further squeezing the margins.

The executives were so intent on winning the argument—and effectively offloading blame onto their colleagues—that Khan felt they'd forgotten about him, not to mention the point of their jobs. Ramazan, a Kazakh member of the leadership who had been silent so far, chimed in with his opinion for why things had gone wrong. He blamed the poor performance on an illogical approach to target setting that broke worldwide sales targets down into regions, then further divided the regional targets among constituent countries. As a result, country teams tried to offload responsibility onto other countries, intentionally underperforming in order to make sure they hit their targets.

After hours of heated back-and-forth among the executives, Ramazan lost it. He pointed his finger at Lars and yelled, "Okay, fine, let me tell you why I wasn't able to make the target last year. It's because of him!"

Lars rose from his seat. "I could have done that order," he yelled back. "Remember those hundred kiloliters we missed because you were not able to deliver it? We were not able to deliver it because your guys didn't send it on time." The bickering continued.

The team's rifts went from the top all the way down. The day before the marathon discussion with the senior executives, Khan sat in on a meeting with the entire sixty-eight-person team for the first time. He was surprised by what he saw and heard. Before the meeting started, the room was a cacophony of different languages—English in one corner, Russian in another, and Arabic in yet another. Members were segregating themselves into subgroups based on native language. Even though everyone spoke English, he noticed varying levels of fluency that exacerbated the division between the groups: native English speakers spoke quickly, seeming to swallow their words, while less fluent speakers remained mostly silent and seemed hesitant to speak up at all. He noticed that the language-based cliques had religious and cultural traditions in common as well.

Khan's head hurt. He wasn't much closer to understanding the root cause of the company's failing revenues, but he suspected it might have something to do with the deep discord in the room. Earlier, he had embarked on a whirlwind tour with Sunil, Lars, and Ramazan of the group's offices in the Middle East and Central and South Asia. One meeting with Farah, a customer service associate, was particularly emblematic of their visits. Farah indicated that he wasn't even aware of the team's drastic drop in performance. What's more, he confided in Khan that he was searching for something to believe in and feel passionate about in his role.

The trip also led Khan to another crucial discovery. While in Uzbekistan, he had gone out to dinner with Lars, a few other team members, and some Kazakh clients. After discussing plans for a new deal, the Kazakh clients had proposed a celebratory toast with vodka—a strong local tradition to mark closing a deal. When Mohammed, a Saudi member of the team, politely declined to drink for religious reasons, Lars told him, "Just do it."

Mohammed didn't respond to this. Lars, in a loud, contemptuous voice added, "Who knows when the Saudis will enter the twenty-first century!"

The table went silent with awkward embarrassment. Mohammed looked down at the table. Khan had heard rumors about Lars mocking local practices during business travel and deriding colleagues' poor English skills, and now he was seeing Lars's cultural insensitivity in action.

Tariq wanted the job, but he wasn't sure whether he could turn around a company whose differences seemed to run so deep.

NEAR, YET FAR—STRANGERS

To understand one of the deep reasons so many of us have so much trouble navigating cultural and linguistic differences, let's go back to 1908, when the pioneering German sociologist Georg Simmel published an essay entitled "The Stranger." In it he asked what happens when a group encounters a person who is similar in some ways to the group's prototype, yet also different. He imagined this person as a traveler, such as a trader entering a close-knit village where everyone knows one another, perhaps for their entire lives. The trader is someone who is near the group in physical space but distant in what might be thought of as social or psychological space. Perhaps the trader dresses differently than those in the village or speaks their language with an accent that marks him as having lived elsewhere. This imagined archetype—a stranger—became an organizing idea around which Simmel's thinking crystallized, grew, and eventually yielded insights that can help us knit together today's most far-flung global groups.

In a sense, Simmel was a stranger-traveler himself. Although he spent most of his life in Berlin, where he was born in 1858, he did not fit comfortably within the rigid academic disciplines of his time and place. As a Jew, he was considered an outsider by German society. In addition, he was part philosopher and part scientist, with a strong connection to the arts. He played piano and violin, married an artist, and wrote about Rembrandt. At the University of Berlin he was a popular speaker whose lectures were attended by crowds of visiting intellectuals. And yet because he was different from most other people in his milieu, many considered him a stranger from whom they felt psychologically distant.

None of his eclecticism or popularity went down particularly well among Berlin's academic establishment. Even after he became a professor, he was denied a chaired position. It was perhaps because he never achieved a top academic rank that he was outshone by some of his better-positioned associates in the burgeoning field of sociology. In fact, Simmel did not become widely known until he was rediscovered in the 1960s by a generation of sociologists who were willing to embrace his interdisciplinary ambitions as well as his metaphorical, almost poetic, writing style. His rediscoverers found that with the passage of time, Simmel's metaphors had become more and more relevant.

Simmel saw that in the modern urban environment, physical proximity inevitably coexists with psychological distance. City dwellers routinely work and live with others who speak different languages (or speak the dominant language in varying degrees of fluency) and follow different cultural norms in their daily lives. In a city, people who pass on the street are unlikely to know each other's families or histories as they would in a small village.

The idea of psychological distance is crucial to understanding global-team dysfunction—as well as remedying it. That's because psychological distance is now recognized as a feature not only of the modern living environment, but of groups of all kinds. It refers to the level of *emotional or cognitive connection* among members. If members understand or empathize with one another, then psychological distance is low, and empathic connections allow for repair of the inevitable fissures. If members don't understand or empathize, then psychological distance is high, and the fissures grow. Globally distributed teams are breeding grounds for high psychological distance.

REDUCING PSYCHOLOGICAL DISTANCE

If the members of Tariq Kahn's team had been collocated, they would have looked at each other innumerable times each day across the table. They would have been forced to see each other's facial expressions and body language and hear one another's asides. They probably would have shared food together at some point. They would have passed each other in the hallway and noticed each other's friendship circles. They might have socialized together. Once in a while they might have overheard one another's family phone calls. Willingly or not, they would have developed multidimensional conceptions of one another that may have been able to accommodate their cultural differences. It is through such many-layered understandings that connections develop, empathy grows, and psychological distance shrinks. You may not love your collocated teammates—or even like them much—but sharing space and time with them encourages formation of an empathic connection.

Global teams are to collocated teams as urban life once was to small village life. When much of the world migrated to working remotely from home due to the COVID-19 pandemic, the new class of remote workers learned what global teams have known for some time: communicating an hour or so over videoconference, no matter how good the technology, is qualitatively different than spending extended time together in an office. Not only do you lose the spontaneous exchanges of impromptu hallway chats and multidimensional knowledge of one another, but because you spend less time in the same physical proximity, your psychological distance increases.

Reducing psychological distance can move the team culture from contention and fragmentation to empathy, respect, and trust. When done right, the geographical distance and national diversity that exemplify global teams can be a source of strength and value that makes one team out of many.

LANGUAGE AS A UNIFYING FORCE

Communication is critical for the functioning of any team. On global teams, where members are rarely native speakers of the same mother tongue, language is often a splitter, creating communication gaps that tend to widen psychological distance. This is an immense topic—I have written an entire book about it—but suffice it to say that the challenge facing teams is to minimize language's divisiveness and restore its unifying power.

In global teams today, English is the common language or lingua franca. At least one in four people in the world now speaks a useful level of English and there are more than one billion fluent speakers. English is considered relatively easy to learn because of its flexible grammar and lack of masculine and feminine forms, but its dominance in business is largely due to the long history of colonial Britain and the status of United States as an international superpower. Although some level of English fluency is typically required to work in a multinational company, the reality is that fluency varies among team members for whom the language is not their mother tongue.

Global team leaders need to manage a number of challenges this variation presents. If English as a common language has been newly adopted into the company as a unifying measure, leaders must realize that is just the first step in the process of organizing a team. Even when English has been established policy within the organization, a lingua franca raises its own challenges for pathways to power and control that differ for native and nonnative speakers.

More specifically, when employees at a meeting break off into subgroups according to a native language that others in the room do not speak, it immediately creates an "us versus them" mentality. As discussed in chapter 2, trust and performance is bound to be low when subgroups operate from this mentality. When the splinter group speaks any language other than English among themselves—no matter how comfortable they may feel speaking Russian, Arabic, Spanish, and so on—they both isolate themselves and exclude others. They are acting much like Simmel's villagers who regard anyone different from themselves as a stranger. Although they may be physically in the meeting room, they are increasing the level of psychological distance.

Native English speakers can create a different set of challenges for the global team manager. Native English speakers' fluency with the lingua franca may lead them to assume more status within the organization than their actual position warrants. By speaking too often, too quickly, unclearly, or using too many idioms and slang words, not only are they insensitive to the struggles that nonnative English speakers can face in communicating, but they also reduce the group's effectiveness. They may mistakenly interpret a nonnative speaker's silence or reluctance to speak to mean a colleague has nothing to contribute. Finally, native speakers risk incorrectly devaluing nonnative speakers' job performance by confusing language fluency with competency.

TWO YEARS LATER: KHAN'S GLOBAL TEAM

To counter the divisive social ramifications of the language dynamics that Khan encountered when he first took the position at Tek, one of his first initiatives in taking the helm at Tek as general manager for the Sales and Marketing Group was to require the entire sixty-eight-member team to enforce Tek's policy of adopting English as the internal business language. Tek was in step with multinational corporations that solve the language barrier dilemma among their diverse employees through a lingua franca. But, in the absence of leaders to equip employees and help them adhere to a set of rules of engagement meant to foster inclusivity, even the best-laid-out language policy can fail. In the early months of assuming his new position, Khan found it useful to hand out the rules of engagement as reminders. He continued to do so periodically, especially when new employees came aboard or if he felt the language dynamics were beginning to fray.

Another early action that Khan did was to fire Lars. It was not an easy decision. True, Lars had a track record of cultural insensitivity, as in his belittling Mohammed for not drinking vodka, and was well known for his impatience with nonnative English speakers who struggled with the language. Lars, whose mother tongue was Swedish, but who had learned English as a child, claimed that if *he* was fluent there was no reason others should not be—completely dismissing the more challenging task of learning a new language as an adult. But Lars was also a skilled veteran employee at Tek, his division distinguished by higher earnings than many of the others. Khan considered trying to work with Lars, to perhaps give him a warning that his behavior had to change, but in the end he decided to act boldly.

The decision to fire Lars sent a message that the team was expected to demonstrate a high level of respect for colleagues from all cultures. Khan felt it was important symbolically and integral to setting a new standard and tone. But he did not stop at symbolic actions. He added "respect for others and their cultural differences" to all employees' annual evaluations, with a higher percent weight given to managers. This was an important move. By instituting cultural sensitivity to the evaluation criteria, Khan not only underscored its importance throughout the organization as a standard to which employees would be held accountable, but also gave himself future leverage should another "Lars situation" surface.

If firing Lars and instituting cultural sensitivity as an evaluative measure was the "stick" meant to punish unproductive behaviors, the "carrot" was Khan's promotion of the idea that diversity was an asset and competitive advantage to the group. He and his team adopted the motto "We are different, yet one."

Khan accomplished this transformation largely by changing the team's culture. After implementing the changes described in "Cross-Cultural Mutual Adaptation" at the end of the chapter, Khan's large, dispersed, diverse team at Tek had lost much of their previous contentiousness and fragmentation and gained mutual understanding and trust. By sending a clear message that their diversity was a competitive strength, he managed to eliminate the pervasive "us versus them" culture he'd found and brought people together via the motto "We are different, yet one." The team's sales grew by 30 percent in two years, market share grew by 6 percent, net profit by 72 percent, and most astoundingly, employee satisfaction went from its nadir of 36 percent up to 89 percent.

Like Khan and his team, global teams can find strength in their differences and expertise. Understanding that performance decrements can quickly take root in global teams is an important first step in ensuring their success. Unchecked psychological distance can be pervasive in global teams where members work across boundaries daily. What's more, challenges and patterns may return as teams add or subtract members, shift or even disband, and regroup. For this reason, the discipline of inclusive communication and mutual adaptation is crucial for teams to first become and then remain aligned. Remote work makes the alignment efforts that much more imperative.

INCLUSIVE CONVERSATIONS IN GLOBAL TEAM MEETINGS

Global teams have to ensure that the fluent English speakers learn to *dial down dominance*, the nonfluent speakers learn to *dial up engagement*, and everyone, especially the managers, learns to *balance for inclusion*.

Fluent speakers dial down dominance. Team members who speak English fluently have to understand the need for everyone to participate fully in discussions and to take conscious steps to include team members who speak less fluently. Leaders have to communicate the fluent speakers' responsibility to change the tone and pace of the discussion by slowing down the speed at which they speak and making a point to use language that everyone can understand. This usually means using fewer idioms or unfamiliar slang terms when addressing the group.

Fluent speakers need to be instructed to refrain from dominating the conversation. Some team members find it useful to limit themselves to a certain number of comments, depending on the pace and subject matter of the meeting. Fluent speakers should also be encouraged to listen actively. Rather than immediately jumping in with their own comment, fluent speakers can first rephrase another's statement for clarification or emphasis. Meeting dynamics are healthy when fluent speakers ask things such as, "I think this is what you are saying?" Likewise, checking to make sure less fluent colleagues understand what had just been said is also very important to create an inclusive environment. Especially after making a particularly difficult or lengthy point, fluent speakers should verbally check in by asking, "Do you understand what I am saying?" These communication behaviors create the conditions that give less fluent speakers the confidence to join the common discourse despite being limited in their language abilities.

Less fluent speakers dial up engagement. Speakers who are less fluent nonnative English speakers have to share the responsibility for the discussion by including themselves. While leaders should empathize with the discomfort some feel speaking English and support language learning opportunities whenever necessary, inviting team members to be heard more often, despite any discomfort, is important. Some nonnative speakers find it useful to monitor the frequency of their responses in a similar fashion as the fluent speakers, but with a different goal: to speak up more. Again, depending on the pace of the meeting, team members attempt to make a certain number of verbal contributions within a certain time frame. As with fluent speakers, nonfluent speakers need to learn to make sure they are accurately heard. Leaders can model asking, "Do you understand what I'm saying?" and push for an honest answer to which they are responsive. Eventually, nonnative speakers will feel comfortable enough to ask a colleague to repeat a point or say it differently if they lose track of a fast-paced conversation. Otherwise, people might nod in agreement even if they did not fully understand what is being said because they would feel embarrassed or ashamed to admit their confusion.

Nonfluent speakers should resist the temptation to speak in their native language when around other team members who would not understand what is being said. Switching between the common business language and one's native tongue is called "code-switching." But code-switching into a language that not everyone knows, and that isn't the official business language of the group, can cause alienation and increase psychological distance on the team. Although the code-switching still occasionally happens in most teams, team members should be quick to apologize when they realize they have slipped into speaking a language foreign to their teammates and translate the conversation for everyone's benefit.

Practice, leaders' encouragement, and knowing that everyone is required, for the good of the team, to follow explicit rules of engagement when speaking make a difference.

Everyone balances for inclusion. Everyone in a team has to take on the necessary role of maintaining balance during formal meetings and informal conversations. Balance means a good mix of speaking and listening on the part of each team member. To a certain extent, team members should track their own behaviors in order to influence this balance. But over time, the goal is that the team develops the norm of tuning in to who is speaking more than listening, or vice versa. Team leaders need to learn to directly ask less fluent speakers for their opinions, proposals, and perspectives. "What do you think?" or "Could we hear your input?" are simple phrases to solicit more participation and subtly intervene to change a group dynamic in a discussion where some have been overly dominant and others reluctant to contribute.

Balanced inclusion for effective group communication is not relegated to multilingual teams. Research has shown that even when everyone speaks the same language, it's crucial that everyone in the group has roughly equal time to speak and listen. Equal participation is necessary for true collaboration—it's how people become engaged with the project or issue at hand. For these reasons, leaders must remind their teams about the need for repeated contributions, and the fact that the nature of global work mandates that each person participate.

Dial Down Dominance	Dial Up Engagement	Balance for Inclusion
✓ Slow down the pace and use familiar language (e.g., fewer idioms). ✓ Refrain from dominating conversation. ✓ Ask: "Do you understand what I am saying?" ✓ Listen actively.	✓ Resist withdrawal or other avoidance behaviors. ✓ Refrain from reverting to your native language. ✓ Ask: "Do you understand what I am saying?" ✓ If you don't understand others, ask them to repeat or explain.	✓ Monitor participants and strive to balance their speaking and listening. ✓ Actively draw contributions from all team members. ✓ Solicit participation from less fluent speakers in particular. ✓ Be prepared to define and interpret content.

Figure 1. Rules of Engagement.

CROSS-CULTURAL MUTUAL ADAPTATION

When it comes to cross-cultural global teams, the old saying, "Give a person a fish and you feed him for a day; teach a person to fish and you feed him for a lifetime," comes to mind. As a global team member, you are bound to come across many interactions—large and small—that require cross-cultural skill and sensitivity. To create "one out of many" requires an ongoing process of understanding and adaptation between team members who originate from diverse cultures and nationalities. To that end, I have derived what I call "the mutual adaptation model."

The model has two interactive cycles—the mutual learning cycle and the mutual teaching cycle. Each cycle helps to slow down interactions and lead to new ways of connecting. There is no special order to these actions; global leaders and their team members find it valuable to engage in both teaching and learning activities during different episodes. Nor are they a one-time set of actions. Instead, they may need to be implemented periodically as reminders. Ideally, employees will integrate these shifts in attitude and behavior to become the norm.

MUTUAL LEARNING

Absorbing and *asking* are the two specific behaviors that comprise mutual learning.

Absorbing. Most people learn by actively watching and listening in on the behaviors of others, similar to the way children first develop cultural know-how as they grow up. As adults, moving out of a comfort zone and into a new context involves similar watching, listening, and "taking it all in." To truly absorb the nuances of a new context requires actively suspending comparisons and deferring judgment. During the absorbing phase, the goal is to gather information about a particular workplace, team, or situation without jumping to inner comments or evaluations. Keeping an open mind is key to understanding different perspectives and alternate practices.

Asking. Learning about a new cultural context can also involve asking questions. The natural give-and-take of one person asking questions and another providing answers establishes mutuality. This very act of give-and-take provides a low-risk, comfortable opportunity for team members to make sense of and adapt to a new context. Asking questions, however, may not always be sufficient to provide a clear or entirely accurate picture; rather, they can serve as additional information and observations that are part of the absorbing phase.

Absorbing and asking are interconnected. Absorbing provides more information and experiences about which to ask questions; asking questions develops a better understanding of observed behaviors. Overall, it's a mindful method that requires leaders to reflect also on their own cultural and national identities.

MUTUAL TEACHING

This second cycle is focused on *instructing* and *facilitating*. Mutual teaching requires everyone on a global team to become both student and teacher. Theories from educational psychology on psychological interdependence that emphasize the central role of peers as coaches and informal teachers make this an ideal process for leaders to introduce to their team members. Cycles of mutual teaching help foster a culture of acceptance among team members, no matter how diverse, and allow everyone to develop multidimensional views of their colleagues and themselves. The collaborative process adds to teammates' increased understanding and appreciation of each other's unique perspectives. The shared experience becomes a common ground among global colleagues, which in turn reduces the barriers caused by psychological distance.

Instructing. Instructing comprises coaching, teaching, mentoring, and other forms of guidance, as well as informal advice and assistance that teammates share with one another in order to help understand new perspectives. In particular, mentoring establishes a personal connection between two or more team members, and frequently occurs between one person who is native to, and another who is new to, a particular environment.

Facilitating. Facilitation is a specific type of teaching behavior. People who facilitate can mediate behaviors and translate cultural meanings among team members. Facilitators are typically familiar with multiple cultural repertoires and can therefore serve as connecting or explanatory links between team members whose backgrounds are markedly different.

The key element to keep in mind about all these teaching behaviors is their mutuality—team members from different backgrounds help, learn from, and ultimately understand one another in the process of becoming part of one united team rather than a disparate set of individuals. All members of a global team should know how to at least instruct and facilitate as a way to build shared learning in the group and further understanding of varying perspectives. Doing so will decrease psychological distance and increase empathy and effectiveness.

In a global distributed team that is likely to have members with diverse cultural and national backgrounds, these cycles of mutual learning and teaching can eventually become habitual, practiced in small moments throughout the workday. As behaviors evolve, teammates build empathy. Mutual learning allows team members to discover, for example, a shared interest in sports or cooking. Mutual teaching allows people to become increasingly comfortable and empathic with one another and less like Simmel's "stranger."

The mutual adaptation model, by providing a way to negotiate and balance our own and others' perceptions of who we are, helps global teams move through the vital process of decreasing psychological distance and building empathy. It's life changing to the extent that individuals become adept at explaining their self and culture to others, can delineate the specific ways they differ from another person, and develop an increased affinity with others both because of and in spite of differences.

Although global teams may include some in-person meetings or interactions, they must, by definition, operate virtually much of the time. In addition to working remotely, people on global teams must also learn to navigate differences that are typically cultural and linguistic. In this sense, they face steeper challenges than do remote teams who share a common culture or language. However, as you will see in the next chapter, differences of many types can divide teams. Even if you all speak the same language and share similar cultural assumptions, you may differ in age, gender, work experience, and training. Some members of your team may be more extroverted and tend to dominate conversations, while other, more introverted team members may hang back and be reluctant to speak. Mutual learning and mutual teaching are not exclusive to global teams; indeed, teams of any sort may want to practice some version of these best practices and key actions to ensure shared learning and to leverage differences for positive outcomes.

Success from Anywhere: Thriving Across Differences

- **Dial it down.** Team members more fluent in a shared language or lingua franca need to slow down the pace of the dialogue and make sure everyone is on the same page. Encourage less fluent members to speak up, and check in with them to make sure they understand.
- **Dial it up.** Those less fluent in the lingua franca need make an active effort to participate in the dialogue despite very understandable fears about speaking up. If you don't catch something, ask colleagues to

repeat themselves. If necessary, monitor the frequency of your contributions and shoot for a quota.

- **Keep the same code.** If you share a native language with some teammates, avoid code-switching between your native tongue and the lingua franca when in a shared virtual space with the whole team. If it happens accidentally, acknowledge that it's potentially inconsiderate, apologize, and repeat yourself in the language that every team member understands.

- **Strike the balance.** Listen as much as you speak, whether it's in a videoconference, email exchange, or group chat. If you notice that certain team members are hesitant to speak up, nudge them with encouragement.

- **Observe and ask.** Step out of your comfort zone, keep an open mind, and absorb what you see and hear from your virtual colleagues. Ask colleagues questions about what you observe.

- **Teach and facilitate.** Proactively share advice, insight, and guidance with your teammates whenever possible. Create opportunities for your teammates to share what they know as well.

- **Empathize.** Allow yourself to feel closer with your teammates through the cycle of learning and teaching.

- **Leverage positive differences.** Divert attention from the differentials that splinter teams, and focus on the diversity of backgrounds that makes your team more capable (not to mention more exciting).

CHAPTER 7

What Do I Really Need to Know About Leading Virtually?

We often say that leaders with strong personalities are "larger than life." The phrase speaks to the leader's outsize presence in the room—the magnetic ability to impress people, captivate their attention, and inspire their respect. This presence is most palpable when holding meetings in the conference room, mentoring employees one-on-one, or walking around the building and checking in with employees through informal chats.

But how does a larger-than-life presence manifest through a computer screen? In my many years of having had the privilege to work with hundreds of virtual leaders around the world, perhaps their most common worry is how to lead without the in-person tool kit that makes them effective in the physical world. Lost is the peripheral vision that tells them who is dialed up with energy at a meeting and who is fiddling with their phone. Lost are the eye contact and body language that allow them to read the ambient mood of the room like a sixth sense. Lost, too, is the opportunity to spontaneously work the room before or after the meeting. Height is literally cut down to the size of a screen. The rich array of sights and sounds that embody the physical world must now be moderated by a single and limited digital channel.

Before we begin to consider the obstacles of a virtual world, let's think about the role of leadership itself. Leadership is an extraordinarily complex task. Leaders must set goals, motivate teams, oversee ongoing activities, avert internal and external constraints, and deliver results. Day to day, week to week, leaders must get everyone on the same page, establish and sustain relationships across individuals and larger groups, ensure team cohesion, and mobilize the team whenever necessary. Add the many more tasks specific to an individual industry, company, and stakeholder, and leadership becomes even more complex.

The introduction of a virtual format can be the straw that breaks the camel's back. Throughout my career, I have witnessed virtual teams collapse. Typically, a company invests considerable resources in convening a distributed group of expert employees for a specific purpose—such as developing a crucial product or honing a strategy—but problems soon emerge. Group dynamics become uncomfortable. Resentments grow. People stop listening.

Ultimately, the work fails to meet the company's expectations. Everyone involved—from the team leader on down—feels the consequences. The breakdown jeopardizes client work, participants' promotions and bonuses, and sometimes even jobs. If you are placed on a high-potential global team that ultimately fails to meet its goals, the company loses clients and money on a global scale. That is the reality.

I have found that for every derailment, there is a manager with a theory about what's happening. Sometimes there are multiple managers with multiple theories and rationalizations, such as: This team member was overbearing. Those members were passive. The assignment was too open-ended—or maybe it was too restrictive. There were too many meetings, or there weren't enough.

I am sometimes asked to get deep into the weeds of these theories and provide custom solutions for each supposed deficiency. Yet no amount of tinkering with the details of personnel, task, or process is going to reduce the overall derailment rate, either within a given company or across the corporate landscape. A closer look traces the problems to leadership. That is where the solution must come.

DEFINING REMOTE WORK LEADERSHIP

As a longtime faculty member and head of Harvard Business School's Leadership and Organizational Behavior course (LEAD, as we affectionately call it), I have engaged with the topic of leadership from every conceivable perspective. I have pursued the how, the what, the when, and the why not only as an academic, but as a practitioner; not only as a teacher, but from the perspective of my students—future leaders—as well. This multidimensional approach has given me insight into the nuances of leadership at every level, from managing individuals through one-on-one relationships to aligning groups around a common vision. Each level of leadership is crucial to the success of a team's performance—and equally difficult to pull off. In the case of remote work, I have adopted a definition of leadership from my colleagues Frances Frei and Anne Morriss: Leadership is empowering other people as a result of your presence—and making sure that impact continues in your absence. Leaders must create the conditions for people to realize their own capacity and power.

Frances and Anne developed this definition of leadership in the early 2000s, when they began working with organizations—including some of the most competitive companies in the world—that were embarking on large-scale change campaigns. They started to notice a consistent pattern among the most successful leaders. Success was not about *them*. Leadership, for the most successful executives, was about setting *other* people up for success. These leaders defined success as creating the conditions their teams needed to thrive. They didn't just hire competent people that they expected to perform well. Instead, their approach was to figure out how to help employees meet their own goals.

What's more, Frances and Anne found that leaders are not only important guides when they are shoulder-to-shoulder in the trenches with their teams; they also stay involved when they are not around, and even when they permanently move on from the team. This discovery is particularly suitable for leading virtually, where so much of leadership has to manifest within the constraints of physical absence.

I have found these constraints are particularly threatening to leaders' awareness of what is really going on. Discord can spread silently throughout a virtual team and elude even the most beloved leaders experienced in managing collocated groups. When teams share a home base, leaders are able to naturally check the pulse of teammates throughout the rhythms of the workday. If an issue is forming, it will become apparent. But without the opportunity to see and hear how teammates are doing, the hairline fractures continue to grow until it's too late and the whole structure snaps. Thus, virtual leaders must understand what it is they *don't* know before deciding on the measures to counter problems as they arise.

In this chapter, I detail the six common challenges that leaders face, how these challenges manifest in a virtual format, and the proven practices to overcome them.

1. Location
2. Class Divide
3. Us versus Them
4. Predictability
5. Performance Feedback
6. Team Engagement

While these challenges arise across teams of all forms, their consequences can be especially harsh in a remote environment. Virtual leaders must proactively search for the earliest warning signs in situations where their collocated counterparts might still thrive under a more reactive approach. If these challenges are left unattended, they will swell into fissures that splinter your remote team.

LOCATION CHALLENGE

The mass migration to remote work in the months immediately following the onset of the pandemic was unique in that everyone was similarly located—at home. Despite differences in home office setups, technology access, or child-care responsibilities, everyone was on more or less equal footing in that they were equally distant from leaders and colleagues; no one was collocated in a physical office.

More common, however, in both the past and in the likely future, is a hybrid structure where some people work remotely most of the time while others might have physical access to their colleagues at least some of the time. Those differences in physical structure are what give rise to complicated group dynamics. Research shows that where people are located in a team or the physical configuration of people on a team profoundly shapes their experience.

Configuration describes not only where people sit in a certain physical location, but also the number of places in which distributed team members are located, the number of employees who work at each site, and the relative balance in numbers of employees at each site. To complicate group dynamics even more, in some cases team members might be working across national borders, which means that people have to also navigate across time zones, national borders, and local organizational cultures.

When teams fail to bridge these boundaries, subgroups form. These cliques or clubs often form around special interests. Distributed teams are also prone to subgroups formed around physical locations. Researchers who study team configurations have found four distinct permutations that drive performance: entirely collocated teams, balanced teams that contain subgroups of equal numbers across two locations, imbalanced teams that contain an unequal number of employees grouped across multiple settings, and teams that contain remote workers—or "geographic isolates"—who work alone in a separate location from other team members. Perhaps surprisingly, members who sit near the headquarters, or with the team leader, tend to ignore the needs and contributions of other people in the team outside their location.

When the scholars analyzed the impact of the configurations, they discovered that members of minority subgroups on teams with imbalanced numbers experienced lower identification with the overall team and less awareness of other team members' expert knowledge compared to their colleagues in the majority. Teams with geographic isolates—either those working from home or single members at a location—were even more likely to feel excluded.

THE CLASS DIVIDE CHALLENGE

People tend to associate power with numbers. Those in the numerical majority group are liable to resent those in the minority group based on the (often inaccurate) belief that the larger group is contributing more than is fair. Solo workers can feel threatened, worrying (again, often mistakenly) that the larger group is attempting to usurp what little power and voice they have. In some cases, these fears are well founded. But even if they're not, virtual leaders must remain attuned to these fears. As a team leader, you will have to help foster equity among particular groups in your team. It's easy to overlook these common tendencies because people don't necessarily express such worries verbally. But regardless of how vocal your teammates might be—or how valid their worries are—these fears ultimately produce the same outcome: inclusive or exclusionary interactions between team members. These behaviors precipitate the performance issues that inevitably derail teams.

Status, defined as a sense of prestige and influence, is another reality of team dynamics that is potentially influenced by subgroup structures and imbalances. Note that the *perception* of status can be as consequential and harmful to team functioning as actual prestige or influence. For example, a study of three international teams in the automotive industry found that Mexican engineering groups perceived themselves as "low status" compared to their Indian and American counterparts. The Mexican engineers, who were used to collaborating closely with one another and asking for help from colleagues, believed (mistakenly) that their international counterparts only valued individual problem-solving and thus feared that their collaborative work patterns would be considered a shortcoming. As a result, they misrepresented their work practices to colleagues in groups they perceived to be "high status," which led to greater conflict and reduced collaboration between the groups. In the same study, when the engineering groups perceived themselves as "high status" as compared to others, they were more likely to communicate openly, ask for help, and share knowledge. Leaders can counter the harmful effects of these distorted perceptions, which include weakened performance and team derailment, by taking ongoing steps to recognize individual strengths in all groups. At the same time, leaders can play down the perceived and real differences in status between members of the team.

THE "US VERSUS THEM" CHALLENGE

Like subterranean cracks that form underneath volcanoes ready to erupt, faultlines are endemic to social groupings of all kinds, dispersed or collocated. Researchers refer to faultlines as invisible or hypothetical distinctions that split a group into subgroups characterized by an "us versus them" mentality. Faultlines form along differences—for example, role functions, expertise, attitude, personality type, gender, age, race, nationality, and language—and create subgroups. A group member may hold similarities to more than one subgroup. For example, in addition to being part of the visible subgroups of people who are your age, gender, and race, you might be one of a few software engineers on the team. These divisions are organic and unavoidable—there is no such thing as a group that is free of faultlines. The question for leaders becomes how to manage subgroups productively before allowing differences to evolve into an "us versus them" dynamic that rapidly erodes team cohesion.

Distributed teams and virtual work add geography as a variable in the mix of potential faultlines. It's easy and even natural to assign "us versus them" terminology to distinguish teams operating in different geographic locations; for example, "the Mexico team" and "the U.S. team." And the more similarity between the members of any one subgroup—for example, everyone is a middle-aged female with marketing experience in the same industry—the more likely it is that this mentality will emerge. The problem is when these faultlines—geographical or otherwise—lead to greater distance from other subgroups. If these natural differences are left unmanaged, conflict increases, coordination issues become harder to solve, and chances for collaboration and productive working relationships diminish.

Faultlines cause trouble when the cracks widen. A few years ago a group of researchers were able to study a Fortune 500 firm's data to see whether they could find a connection between faultlines and team performance. They examined the records of dozens of teams comprising more than five hundred people engaged in complicated, nonroutine, highly variable tasks—in other words, typical knowledge work in a big company. They looked at the social distinctions of gender and age as well as informational distinctions of education level and company tenure, focusing on two characteristics of faultlines: *strength*, or how cleanly a subgroup was divided from the rest of the team; and *distance*, or the size of the gaps. To understand *strong* faultlines, think of a four-person team on which two people were young and male and two people were older and female—the age and gender subgroupings would align perfectly, and there would be only one clean way to subdivide the team along these attributes. If the age gap was wide—for example, the two young males in their twenties, the two older females in their sixties—then the faultline *distance* would be considered large.

To get a handle on teams' ability to meet goals, the scholars dug into the data to look at various types of bonuses that had been handed out to reward each group's performance. They also looked at evaluations, including those from employees, and used quantitative methods to analyze the prevalence of key words and phrases. Using these measurements, it turned out that greater faultline strength around social distinctions was associated with poorer team performance, with distance exacerbating the effect. On Tariq Khan's team at Tek, which you read about in chapter 6, the faultlines separating the participants who were based in different countries were strong, and the groups were distant. It is not surprising that the team's performance was in precipitous decline at a key period.

My study with two colleagues took an in-depth look at faultlines on software development teams. We interviewed and observed ninety-six global team members at a Germany-based software company. Because we were interested in understanding subgroup dynamics in teams spread across locations, we simultaneously observed team members located at multiple locations. This format allowed us to record social interactions and team dynamics as they occurred and provided rich data on how individuals at each of the two locations experienced team interactions. The approach also allowed us to observe how people experienced cross-site and local meetings, and how events were interpreted similarly or differently across sites. We paid close attention to the interactions, attitudes, and responses of individuals as they communicated with collocated and distributed colleagues alike. We also attended meetings, observed conference calls, had lunch with people, and went to after-work social gatherings.

We found that faultlines around English language fluency and nationality created strong divisions and engendered an "us versus them" in some but not all of the teams. What determines which teams are most susceptible? The data we compiled suggests that divisive subgroup dynamics occurred only in teams that also suffered from power contests; in other words, power contests activate otherwise dormant faultlines. When strong negative emotions radiate tension across locations they can trigger a self-reinforcing cycle that fuels an "us versus them" dynamic. As this negative feedback loop spiraled within the software development teams we studied, resentments grew among team members. People started to withhold information from their distant colleagues. Group performance declined. In the worst-case scenario, teams disbanded.

Perhaps most important, we discovered that team leaders were often unaware of the underlying issues that activated faultlines and caused dysfunctions in their team. They could sense that something was wrong, but often didn't know what and why.

The problem is that faultlines tend to harden into solid boundaries, leading subgroups to compete with one another. Employees begin to stereotype one another, and subgroups begin to define themselves as superior to other subgroups. At Tek, the treatment of certain team members as second-class participants, as well as Lars's deeply offensive and racist comment about his Saudi colleague, were classic behaviors that created an in-group and an out-group and resulted in members acting as though they were on separate teams.

However, faultlines are not always all bad. Researchers have found that under certain conditions, faultlines around education levels and tenure had no harmful effect on team performance in the company they studied, and such divisions can even promote effective decision-making. Groups often thrive on faultlines, feeling energized by subgroups' divergent perspectives or expertise. The problem is that such thriving can turn into a "we love to hate such-and-such a group" energy that is eventually limiting and insular. The challenge facing leaders is to be aware of these dynamics and find a simple overarching strategy for repairing the cracks among dispersed team members. Well-led teams tend to be resilient. Leaders can help teams draw strength from diversity in members' expertise, personalities, and rich backgrounds.

How can leaders help teams bounce back from faultlines? One way is to help teams redirect their fractures. In many cases, team members engage in what is known as "reappraisal," essentially directing themselves to be positive and more empathic toward other participants.

Team leaders, knowing people's tendencies to act certain ways in groups, can counteract faultlines by playing up some group aspects and downplaying others. First, build and stress one *group-level identity*: the umbrella identity that binds the team together into one, rather than a fragmented, entity. Remind team members that they each represent the team (for example, marketing or design team). Second, emphasize *superordinate goals*: the higher and common purpose that team members are trying to achieve for the firm. Remind them that each person, no matter their background, will help the team reach that goal. When concerns do arise based on perceptions of skewed power, choose to deal with them selectively. Sometimes it's best to redirect the focus of team members from power squabbles toward the greater goal of innovating to help society, increasing revenue, or beating the competition.

THE PREDICTABILITY CHALLENGE

Virtual leadership requires frequent communication with team members. Hearing from the boss helps make the present and future more predictable. Such predictability gives shape to daily work. When communication cannot be in person, the various digital tools discussed in chapter 4 are all crucial in establishing a virtual presence. An increase in communication from the team leader that is clear and direct can accentuate the positive effects of remote work and compensate for the negative. As I noted in chapter 2, remote team leaders play a role in the degree of professional isolation that remote workers do or do not feel at home. Setting clear goals and providing involved feedback are always integral to good leadership. While such basic management is relevant for collocated teams, it is even more important when the team members are out of sight and therefore removed from whatever natural flows of communication take place in the office.

Indeed, research confirms that when leaders increase communication regarding job responsibilities, expectations, goals, objectives, and deadlines, employees are more loyal to their companies, more satisfied with their jobs, and performed better. Team members also respond favorably to leaders who implement online social groups and provided regular reviews about work performance, salary, and career development.

THE PERFORMANCE FEEDBACK CHALLENGE

Leaders must routinely provide feedback to remote teams to ensure positive results as well as weigh in on individual performance reviews and promotions. One question that is always on the minds of virtual team members, particularly those who are isolates or "out of sight," is whether they fare differently than collocated workers who can physically catch the boss's eye and ear via an in-person relationship for an informal "job well done," or "needs more effort." Researchers who wanted to know the answer to this question surveyed a group of supervisors who were responsible for both remote and in-office workers, as well as their direct reports.

Although the researchers, much like the workers themselves, had feared that remote workers would be evaluated more harshly or given lower performance reviews than their in-office peers, in fact they found that working remotely did not have a negative impact on either the relationship or task dimensions of job performance evaluations. Also, each individual's career advancement prospects were measured by their supervisor's response to the question, "How would you assess the employee's chance for advancement" = good chance, very good chance. Supervisors' evaluations of career prospects for remote workers did not significantly differ from those of nonremote workers. Out of sight did not mean out of mind.

THE ENGAGEMENT CHALLENGE

One of the most important "tools" for leading virtually is having an influential process that enables you to engage consistently with team members when you are not physically present. By process, I mean the behaviors and interactions that you practice over time, even if they seem trivial. If you are unable to "read a room" or walk around a building to interact with employees, you have to deliberately create moments for self- and other-awareness by actively soliciting insights about group and company improvement. Encouraging people to express observations is important for remote workers to fully engage in their virtual team's success. To create the conditions for people to realize their own capacity and power, a leader must attend to the traditional, collocated aspects of team processes as well as initiate those that are unique to leading virtually. To this end, I have learned that three common practices are important for leaders to implement: 1) structuring unstructured time for informal interactions, 2) emphasizing individual differences, and 3) forcing conflict.

Leaders need to structure unstructured time to promote informal interactions. Fostering a more laid-back and informal atmosphere in a remote format requires intentionality in much the same way that leaders who are adept at nurturing collocated teamwork often move members' workstations or offices into close proximity. That is because of the well-known benefits of informal interactions about things outside the team's official work: topics such as the weather, families, sports, a new restaurant, or a TV show. Conversations like these build relationships and give members the sense that they are being heard. At the same time, informal chat may provide valuable nuggets of work-related information as members talk about their experiences. A team member's offhand complaint about phone system glitches may uncover a serious technical challenge that needs to be addressed. A member who follows local politics may mention proposed legislation that would affect the company's bidding process.

Casual, spontaneous communication is rare on distributed teams, which tend to convene for specific tasks and in contexts that put time at a premium. Leaders must therefore make a conscious effort to promote spontaneous interaction. A simple intervention is to set aside the initial six to seven minutes of a meeting for informal chat about nonwork matters. Members should be encouraged not only to talk about the weather but also to communicate—and, yes, complain—about constraints such as technical and work conditions. Leaders can also facilitate informal contact by scheduling virtual lunches; breaks for coffee, tea, or a snack; and even a virtual happy hour. Teams can also come up with plans for virtual recreational activities that they vary from time to time to hold their interest.

Leaders should demonstrate the value of informal talk by initiating it themselves. After a manager inherited a remote team as part of an acquisition, he made a point not only to involve those virtual employees in important decisions, contact them frequently to discuss ongoing projects, and thank them for good work; he also called team members personally to give them their birthdays off and to simply chat. However, leaders do not have to be present for every such instance; in fact, it's a good idea to also facilitate unstructured peer-to-peer time. Leaders can pair people up to check on each other regularly, at least once a week, for a virtual activity. Each person can be asked to do something that expresses gratitude related to their coworker, for example a gift card, something fun for their loved ones, or sending a handwritten note. This shapes familiarity, bonding, and connections outside work and breaks the isolation for everyone involved. Rotating the peer-to-peer pairings among the team means that each person gets to connect and bond all over again, this time with a different person.

Another important practice for leaders—emphasizing individual differences—keeps the team abreast of peer strengths they can leverage. Unless a leader actively encourages differences of opinion, members are too often hesitant to voice their point of view. It's easy for leaders to place so much emphasis on organization and efficiency that they inadvertently quash the expression of divergent viewpoints, even from members with deep expertise. In one case I studied, a software developer was a member of a team whose leader brooked no dissent, so to protect his position he stayed mum and didn't express his disagreement with the design of a particular feature. Four weeks later, the team became ensnarled in the problem he had foreseen.

To promote a free exchange of views, leaders should ask others for their opinions: "What do you think about the new proposal?" "Does anyone have additional comments?" Agenda items, too, should be open for discussion. Emphasizing such differences also highlights individuality while downplaying subgroup boundaries. Leaders should avoid referring to people by their membership in a subgroup ("As one of the New York guys mentioned . . . ," "As one of the engineers said . . .") and should instead focus on the perspectives and knowledge of individuals.

Forcing productive conflict about ideas, tasks, and processes will strengthen the group's work and is central to creating the conditions for people to realize their own capacity and power.

In remote environments conflicts or disagreements are even less likely to occur organically and consistently than they are in collocated offices. Ideally, leaders and members should feel psychologically safe enough to tease out conflicts, treating them as opportunities to learn. To this end, teams should frame dissent as positive—as differences of viewpoint—and provide assurance that members will not be blamed for "rocking the boat." Divergent opinions should be met with comments such as "I like that idea . . . let's brainstorm more like it." If others are dismissive, their comments should be channeled into specifics: What are their concerns? That way the proponent of an idea can take an active role in shaping the discussion by addressing others' questions. If such a gentle approach is not working, leaders should force conflict into the open. This does not mean inviting team members to vent their grievances or harp on personal and cultural differences; it means intentionally eliciting open intellectual disagreement that can spur innovative thinking about a given task or a process.

When you lead virtually, you lose face-to-face contact and the in-person tool kit that made you so effective in the physical world. All of your hard-earned gains, built on the foundation of your leadership presence, fade to the background. The sights and sounds that embodied the world for you are moderated by a single and limited digital channel. Serendipitous and planned informal encounters are absent. You can't drop in on someone to invite them for a coffee break, nor can you take your team members to lunch to swap stories in order to strengthen your bond. Despite these losses, virtual leaders can still equip and empower their team. Your goal is to make sure the impact of your leadership work continues in your absence by creating the conditions for people to realize their own capacity and power.

Leading virtually, though multidimensional and uniquely challenging, can be rewarding. Much of the time it's learning to reorient yourself from your in-person tool kit that relies on physical presence and informal communication to virtual equivalents or entirely new tools. Many of the rules for leading in collocated conditions still apply, but for remote teams you have to be more mindful and conscious in your efforts to achieve the same results. Leading virtually often requires you to be more formal to make interactions feel informal and more structured to create open time for informality. Understanding the various ways that subgroups and faultlines can form when people work in distributed teams—and discouraging the inherent divisiveness—is key. Equally important is making sure to communicate regularly and consistently to remote team members who are not visible. Once you become attuned to the inherent risks of remote work and establish the necessary countering measures, you will enjoy a loyal remote group that performs in accordance with each member's unique capacities. You and your team will feel empowered to handle any situation that might arise.

Success from Anywhere: Leading Virtually

- **Minimize differences.** Where people are located matters. Differences in distributed team members' geography, as well as differences between team members who do or do not work remotely, can engender subgroups and social dynamics that result in conflict. Leaders have to become aware of and actively manage these differences, especially for isolates.
- **Emphasize strengths, not status.** Class divides will form among groups based on differences in size along with real or perceived differences in status. Leaders can counter the harmful effects of perceptions about low status by taking ongoing steps to recognize individual strengths in all groups and by downplaying perceived and real differences in status between members of the team.
- **Promote a common purpose.** Faultlines will develop in every team. Leaders can work against faultlines' corrosiveness by building and stressing one *group-level identity*: the umbrella identity that binds the team together into one and reminding team members that they each represent the team. Leaders can also emphasize *superordinate goals*—the common purpose that team members are trying to achieve—by reminding them that each individual effort contributes to the team goal.

- **Create structure.** Remote workers crave predictability. Leaders can support this by providing clear, consistent, and direct communication about job description and responsibilities.
- **Give feedback.** Remote workers are no less likely to perform well and advance in their careers than collocated coworkers. Leaders must provide appropriate and constructive feedback to support individual goals.
- **Promote engagement but don't avoid conflict.** Making sure your team gels is a ceaseless endeavor. Creating structured unstructured time for informal conversation at the beginning of virtual meetings as well as virtual fun times can help teams to bond. Leaders can also encourage team members to appreciate each other's differences and make it safe enough to voice disagreements or concerns.

CHAPTER 8

How Do I Prepare My Team for Global Crises?

Trouble was brewing in Istanbul. Antigovernment demonstrators raged in the city's beloved Taksim Gezi Park, making international headlines. Riot police, attempting to gain control, shot canisters of tear gas into crowds of pedestrians. Istanbul is divided by the Bosphorus River into two continents, Europe on one side and Asia on the other, its connecting bridges a source of pride. But now, in the sweltering summer of 2013, a new divide was becoming entrenched in Turkish society that was nearly impossible to bridge—a generational disagreement between the forces of progress and tradition. As the crisis deepened, civil unrest took over the country. Mainstream Turkish society spouted long-standing anti-American rhetoric. When protesters poured out Coke drinks in the streets, vowing never to consume Coca-Cola products again, the rhetoric turned into action. The Coca-Cola brand, seen as an iconic American product, had become the symbol of the West's interference and oppression in Turkey.

Galya Molinas, president of Coca-Cola's Turkey, Caucasus, and Central Asia Business Unit, was keenly attuned to what the public spilling of soft drinks might mean. Her predominantly female senior team had just enjoyed a seventeen-month run of record-breaking performance and volume growth. A twenty-year veteran of the company with a stellar record—known for a warm, friendly smile and a demeanor that exuded competence—Molinas was a paragon of the modern Turkish leader. But along with other American firms, sales in her business unit had already plummeted dramatically in response to the political unrest. She was well aware that her team's unprecedented success was now threatened by external events beyond her control.

Molinas and her team serve as a compelling example that in today's increasingly global business environment, what occurs in one country or region of the world can ripple globally. If you are leading a global team, chances are you will encounter an equivalent experience to "people spilling soft drinks in the public square" that is motivated by external events. The interconnected world markets in which we all live and work are major factors in creating a continual series of smaller crises and in requiring that all leadership be global. Any lingering doubts that the world is profoundly interconnected were laid to rest by the 2020 COVID-19 pandemic, which forced a global crisis that wreaked havoc not only in the lives of millions of people who were suddenly forced to work remotely but also in geopolitical relations as countries alternately cooperated and competed for resources.

In the last century, the prominent Massachusetts politician Tip O'Neill coined and made famous the notion that "all politics is local." He believed that whatever happened in small communities was consequential for larger government, and that effective politicians remained in touch with the everyday concerns of their local constituents. Today, it's more accurate to also say that "all leadership is global." Understanding how global issues impact local sensitivities is key. No matter how local your domain, you must remain in touch with current global issues and learn to develop new capacities in response to globally induced crises that may affect the organization. Globality and locality have to function together.

In this chapter, I will first explain how increasingly volatile, uncertain, complex, and ambiguous (VUCA) factors create the conditions for the worldwide ripple effects of this interconnectedness and characterize a world where crisis is to be expected. Next, I introduce the mind-sets that leaders of global teams must cultivate and develop to navigate our interconnected, crisis-prone world and explain what that has to do with an iPhone camera. You will learn about what sociologists call *the country-of-origin effect* and why it matters for global companies, as well as why a cognitively diverse team is essential for the challenges presented by an interconnected world. Throughout the chapter you will follow the story of how Molinas and her team learned to successfully meet the crisis of their plummeting performance and then about Molinas's leadership to meet the revolutionary changes that COVID-19 brought to the world.

VUCA: THE WAY WE LIVE NOW

In an era of globalization, companies must function in a world with VUCA. Adapted from a term used by the U.S. Army War College to describe the environment that military leaders must operate within, VUCA today can apply to conditions such as a market crash, a natural disaster, or a public health crisis. VUCA is a constant in the world in which we now live. Although what follows may sound familiar, having the vocabulary and a fuller comprehension of the way the world currently exists is a first step in figuring out how you can best meet its considerable challenges. The examples I provide are the tip of the iceberg; no doubt you can supply plenty more.

Volatility describes a state of constant change that is dynamic, sudden, and rapid.

Protesters pouring Coca-Cola into the streets were an unexpected challenge to the company; no one could have predicted what form the protests would take and how long they would last. Other examples of volatility include fluctuating prices following supply shortages after a natural disaster or the rise and fall of COVID-19 infection rates around the world. Earthquakes and floods create physically volatile conditions for disaster relief workers.

Uncertainty refers to the unpredictability of these sudden and rapid changes, making it difficult to anticipate events and prepare accordingly.

Molinas understood that the political unrest affected Coca-Cola's sales, and though she would eventually act to remedy the situation, she could not be certain that the changes she made would raise revenue. Other, more general examples of uncertainty include the market entry of new products from a given company's competition or the timing and effectiveness of a new vaccine. Hiring freezes, unemployment numbers that rise or fall, or the effect of new government regulations create uncertain conditions in which leaders must operate without being able to accurately predict outcomes. Uncertainty often involves more than one factor, like the market entry of a new product in a country with new regulations *and* a distressed economy.

Complexity involves situations that have many dimensions and moving parts whose sheer volume creates conditions that are difficult if not impossible to control.

In 2013, Molinas was at the helm of a territory that included eight countries in Central Asia in addition to Turkey; any changes she and her team made would have to accommodate a plethora of decentralized local teams, bottlers, markets, consumers, regional leaders, and cultural traditions, as well as decisions and dynamics of the Istanbul headquarters. Multinationals are by definition organizations that operate within a web of complexity, which includes the laws, regulations, and customs of various countries. Complexity is also a given for many organizations in today's society. Hospitals, financial institutions, technology hubs, and airports operate with degrees of complexity unimaginable to anyone who lived a century ago. Complexity is a guaranteed setup for things to break or go wrong in ways that lead to crises large and small.

Ambiguity refers to situations in which one faces "unknown unknowns" and causal relationships are unclear.

In 2013, Molinas could not know how decisions made by leadership in Istanbul would affect local markets in Azerbaijan or Uzbekistan, two of the countries under her jurisdiction. Similarly, the local market details in Armenia and Kazakhstan, also part of her Central Asian territory, were by definition unknown unknowns because of their geographic distance to Istanbul. Like other global leaders entering an emergent market, she was bound to work in conditions rife with ambiguity. Seven years later, she would work in even greater ambiguity—as well as volatility, uncertainty, and complexity—when the worldwide pandemic caused by COVID-19 blew open the world. I will describe how she responded a little later. We don't know what the long-term effects of COVID-19 will be on organizations, industries, and societies, although we do know that the world has profoundly changed. During the pandemic, every government leader had to make decisions that weigh the losses and gains of quarantine versus business as usual while not being able to accurately predict the consequences of whatever strategy they designate.

Taken together, the volatility, uncertainty, complexity, and ambiguity that make up the world in which today's business leaders must function is a powder keg for periodic crisis. Market crash, natural disaster, public health crisis, political upheaval—any and all are unexpected crises that global leaders must prepare for and expect.

In a fundamental sense, to prepare your team for a crisis means extending your concern far beyond your own team, markets, or industries. In my years of meeting and speaking with hundreds of leaders from around the globe about how to best prepare to meet such crises, I have learned that team leaders, whether they serve mostly international or local markets, must develop Global Leadership Aptitude. This aptitude demands that you learn to *develop panoramic awareness*, actively *frame the situation*, and exercise the capacity to *act immediately*. Each of these skills has broad applications and interpretations. In the rest of this chapter we will follow how Galya Molinas used these three skills.

DEVELOP PANORAMIC AWARENESS

To understand panoramic awareness, think of a camera lens, especially the one so many of us now use on the ubiquitous iPhone. We know to use the camera's landscape lens to photograph a wide swath of countryside or a 360-degree view of a room. To snap a close-up picture of a single tree in a landscape or a friend's face in a room, we use the portrait lens. In much the same way, global leaders must learn to shift their attention from a wide swath of events, which are often international in scope and involve crisis, to close-ups of, for example, team dynamics or local sales figures.

Scanning current global issues is the first step in developing panoramic awareness. Leaders don't have the luxury of consuming news solely in one part of the world. Like the landscape lens, you must constantly maintain as wide a view as possible on international events, including the fluctuation of oil prices, regulatory or labor law changes, and shortages or surpluses in agriculture that could impact entire ecosystems. Whether they are transient and fast-moving or changeable, it is important to be vigilant and investigate the relevance of global events. One simple yet essential practice is to consume a variety of international media on a consistent basis. That will allow you to better grasp events, geopolitical or otherwise, which is a first step for defining the local problems your "portrait lens" perspective will pick up in your own business jurisdictions.

Recently, I asked Molinas what media sources she follows to stay informed. She confessed that she does not watch the daily news on TV because it does not "provide insights or understanding," but is instead "highly politicized." Instead, she listed the media she reads online: BBC, the *New York Times*, *Wall Street Journal*, *Financial Times*, the *Economist*, *Al Jazeera*, and the *Atlantic*. She told me that she was also "fed" by the well-read and cognitively diverse people with whom she is surrounded; their interests in such varied fields as biology, politics, medicine, and sociology provide knowledge sharing at a high and comprehensive level.

Keeping vigil over a volatile global landscape, in which lightning could strike at any moment, can often come at the cost of a good night's sleep. The anti-American sentiments that culminated in pouring Coca-Cola into the streets of Istanbul initially shocked and paralyzed Molinas and her entire high-performing senior team. Months earlier, their numbers had broken records! Their business unit had been the only country within the entire company that had delivered double-digit volume growth for two consecutive years, winning a dozen Coca-Cola awards in 2010 and 2011. Business teams from China had visited, wanting to learn how Molinas and her team had successfully met the challenge of working with many local bottlers distributed across a large geographical region. Now she was watching a dangerous decline in revenues with no clear sign of when and how the hemorrhaging would stop. The trouble brewing in Istanbul, a classic VUCA situation, threatened to put an abrupt and premature ceiling on Molinas's success, even crippling the team.

When she scanned her lens into panoramic awareness, what dominated the landscape at the time was the anti-American sentiment rampant in Turkey. Social scientists call this phenomenon the *country-of-origin effect*. Understanding the country-of-origin effect and its consequences will help you to identify and handle one of the most common crises that you may face as a global leader in an interconnected world.

GRAPPLING WITH THE COUNTRY-OF-ORIGIN EFFECT

The country-of-origin effect, a term first coined by sociologist Robert Schooler in the mid-1960s, has tremendous consequences in a global economy, especially for marketing, making it a top priority for leaders. Simply put, the country-of-origin effect comes into play when consumers stereotype a product or a service according to preconceptions about the product's country of origin rather than its intrinsic value. A country-of-origin effect can be positive, but more often it is negative, as in Turkey's rejection of Coca-Cola because of citizens' anti-American sentiment.

Global leaders must anticipate the threats that country-of-origin effects pose to revenue performance. These threats can take the form of boycotts at an unprecedented scale. Today, social media networks connect a global, remote network of consumers who can communicate information instantaneously, organize large-scale campaigns, and mobilize against companies headquartered in countries whose politics they find objectionable.

Let's look at some examples of country-of-origin effect in one area of the world that has required leaders to develop heightened levels of panoramic awareness—the Middle East. Motivated by anti-Western sentiments, a number of consumer boycotts and protests have reflected the region's volatile political upheavals and changes.

A consumer boycott forced Sainsbury's, the leading British supermarket chain, to exit the Egyptian market in 2001 after the multinational company had incurred more than $125 million in losses in two years. Although they provided jobs and popular products, negative sentiment against the supermarket chain had erupted in Egypt from stories about the company's supposed links to Israel. Consumers used the boycott to voice their objection to the Israeli military's response to protest riots in the Palestinian territories. In another example, Arla Foods, a Danish company long established in the Middle East, nearly had to withdraw from the entire Middle Eastern market after a 2006 boycott following cartoons published by a Danish newspaper were interpreted as mocking the Islamic faith. Other than the fact that both the food company and the newspaper originated in Denmark, no clear link existed between the two organizations; nevertheless, Arla had to respond.

Country-of-origin effects can rely more on perception than on fact. Effective communications and messaging are key in converting or reframing intensely negative perceptions into positive reasons for consumers. Arla was able to remain in the region by proactively disassociating itself with and publicly denouncing the offending cartoons in full-page newspaper ads throughout the Middle East. Similarly, when rumors circulated that Nestlé, a Swiss multinational, was using milk powder from Denmark, Nestlé resorted to printing ads in Saudi Arabian newspapers to inform consumers that its products were *not* of Danish origin.

Global leaders need to understand and become aware that consumer boycotts are often linked to international political events rather than corporate practices. Not too long ago, when I discussed country-of-origin effect with Mexican CEOs, the topic of the 2016 U.S.

presidential elections came up. The majority of the CEOs in the room said that they were paralyzed with shock and fear once Donald Trump was elected as the forty-fifth president of the United States. As a candidate, he had vowed harsh trade rules for Mexico, increased deportations, and campaigned with threats of erecting a U.S. Mexico border wall. Soon after the election, Mexican consumers began boycotting American products. Specific calls to boycott U.S. companies in Mexico included McDonald's, Walmart, Coca-Cola, and Starbucks, with heavily trending hashtags on social media in Mexico such as "#AdiosStarbucks," or "Goodbye Starbucks." In addition, supplier relationships were being severed. Bids for contracts that had been easy to capture were fleeting. The impact to the Mexican peso hit every company hard.

American consumers are not immune to the country-of-origin effect. When social scientists surveyed five hundred randomly selected residents in Texas about their willingness to buy products from thirty-six test countries, which had a range of socioeconomic and political structures, they found that participants "were most willing to buy products from economically developed free countries with a European, Australian or New Zealand culture." In other words, people are most likely to prefer products from countries with belief systems and cultural climates that resemble their own, and least likely to want to buy from those they perceive as antagonistic or dissimilar.

FRAME THE SITUATION

When you use a camera to snap a picture in portrait mode—a family portrait, a well-prepared meal, or even a selfie—you choose how to frame the image. How much of the background to include? From which angle? How many clicks before you get it right? Framing the situation to prepare your team for a globally induced crisis works in a similar fashion. Once you have scanned a panoramic landscape and can anticipate risk that lies ahead, you must take a close look at changes you might make in your team to meet the future challenges wrought by global events.

For example, in the months leading up to the 2016 U.S. election, Mexican CEOs might have better anticipated multiple scenarios to counter any effects to their products or services. Framing the situation, by moving your lens from panoramic to a series of framed portrait scenarios, allows you to actively anticipate how current events might affect both the short-term and long-term future. As the global political climate becomes more polarized and customers correspondingly react to that polarization, actively anticipating and framing potential scenarios becomes crucial.

A kind of litmus test for competent framing of the situation was apparent in the varying responses leaders adopted in late 2019 and early 2020, when the world first became aware of the existence and threat of a novel coronavirus spreading rapidly across national borders. Consider, first, how the leadership in New Orleans, Louisiana, framed the situation.

In the month or so leading up to the annual Mardi Gras festival, which attracts more than a million people from all over the world, the U.S. government issued a report that the novel coronavirus detected in China was a low threat to the American public. The mayor, along with the city's top health official and the carnival planners, took the report at face value. That miscalculation had tragic consequences. Mardi Gras was held on February 25, and over 1.4 million people in closely packed crowds reveled in the city streets. February 25, 2020, was also the day that the U.S. Centers for Disease Control issued clear warnings about possible disease spread and that cities should begin planning strict measures for its containment. But it was too late. The damage had already been done. The situation had been framed incorrectly, or out of focus, or not at all.

Nearly two weeks later, on March 9, the first coronavirus patient was identified in New Orleans. The virus spread unabated, and the city soon had one of the fastest-growing rates of infection and mortality to date in the United States. Afterward, the mayor defended the decision to hold the festival, insisting that the city had acted in accordance with the information that they had been given at the time; the federal government had not issued any warnings about the United States being "potentially on the verge of having a crisis for the pandemic." Yet the news that a deadly virus was on the loose was already out. Apparently, it's not enough to spot a looming crisis on the horizon; leaders must also accept its reality. Experts have found that when leaders don't heed warning signals, it's often because they mistakenly believe their organization to be invulnerable or because by denying the reality they can hold on to a sense of normalcy. The leadership in many other cities across the United States and the world had framed the situation from a number of angles, most likely consulted a range of sources, recognized the approaching threat, and taken whatever necessary precautionary measures they could. New Orleans either could not or would not frame the storm that was on the horizon and suffered severely for it.

Singapore, by comparison, was exemplary in how they framed the situation when they saw the potential lethality of a novel coronavirus coming out of Wuhan, China. Like some other countries, such as Iceland, New Zealand, and South Korea, Singapore was at first very successful in containing the number of infections in its population and kept their numbers low. The government health care system was able to ward off the disease threat by mobilizing significant resources ahead of time. Experiences with other public health crises, such as the

SARS outbreak in 2003, informed the country's leaders as they clicked into portrait mode and anticipated various scenarios that might befall their population. Accurate and effective communication channels between official agencies, medical administrators, and personnel were already active; testing and tracing technology was in place; and people understood the purpose of stay-at-home precautions. The population remained relatively unscathed because leadership at many levels was able to accurately frame and anticipate the 2020 COVID-19 health crisis by building a coordinated, country-wide task force. Singapore's leaders clicked into and framed the risks they saw on the horizon.

GENERATE SOLUTIONS WITH DIVERSE MINDS

The crisis unfolding in 2013 around Molinas in Turkey, while not as volatile, complex, or far-reaching as the COVID-19 pandemic (although she too would have to reckon with the events of 2020), is more indicative of the crises that global leaders face on a regular basis. Her multi-country business unit was failing, in large part due to changes wrought by the shock of external events. A situation framing was needed. The first place Molinas trained her close-up portrait lens was closest to home—her own team.

Molinas described her team as "great analysts, great marketers, great people." All but one was a woman in her forties, and all had similar professional and cultural backgrounds. They seemed to work well together—even easily. When an issue came up, everyone agreed with one another's opinions. There wasn't much discussion; they were, as Molinas said, "very nice to each other and all looking in the same direction." No one challenged one another's perspective, and Molinas had sensed that the unanimous agreement did not bode well for the future of the business. But the team had been winning awards for rising revenue, so she didn't rock the boat.

However, when the crisis hit in 2013, she had to accept that her excellent team was a liability. Despite their combined years of experience, she saw that her team was simply not equipped to reinvent a business in response to a crisis wrought by protests and anti-American sentiment. In order to actively anticipate whatever might come next and then make decisions for how to change course, she needed a team that could approach issues and problems in new ways. Essentially, she had to reframe the situation.

Practically speaking, the team did not have any prior experience with emerging markets to understand, interpret, and handle the political and operational risks with their consumers outside of Turkey in the emerging markets of Central Asia. Most crucially, there was no one who could bring insights and solutions to a business struggling due to unease in society at large. Molinas knew she needed people who came from diverse environments and who had similar experiences elsewhere if she was to deeply understand the dynamics of emerging markets and start planning in advance for how to remedy the situation.

These insights did not come all at once. Molinas told me she anguished for days and days trying to figure out the best path forward. She needed to stop the bleeding, and, under the conditions, her team had nothing to offer. Instead of adopting a reactive or defensive response to the crisis—as some leaders might have done—she genuinely tried to understand the problem at its roots and think through what management changes she would make. Finally, when she had framed the situation from enough different angles, could clearly see the challenge ahead, her leadership came through in a bold declaration, "We are going to get the best available talent globally for all critical positions." In other words, she'd realized that having business experience in unpredictable economies and emerging markets became more important than any other qualification.

"There was a certain uneasiness and tension among the leadership members, but we did what we had to do, and some brave decisions had to be made," Molinas explained. "We hired more men and people from other countries." Reflecting on what the experience taught her, she explained, "I have learned that diverse thinking is crucial for global leaders. You cannot survive just by being a great economist or a great finance person. You need a thorough understanding of the various countries' dynamics. For that, you need to have a strong, diverse, and experienced team."

Molinas's insights align with what people who study optimal team composition also know. Global teams whose members reflect a breadth of demographic backgrounds, genders, religions, and cultures are better equipped cognitively to find effective solutions that address change. The combination of unique perspectives and experiences is more likely to bring unique points of view to the table, and ultimately result in better problem-solving.

Three seasoned executives—one from Mexico, one from South Africa, and another from Greece—joined the leadership team in 2015 and 2016. Molinas reflected: "They brought in experience from more than twenty emerging markets in Asia, Russia, the Middle East, South Europe, Africa, and Latin America. The HR director, from South Africa, worked in Namibia and throughout sub-Saharan Africa. The marketing person lived in Venezuela, where challenges were similar to those in Uzbekistan." She made sure that each new hire had relevant experience from more than one country and had faced similar conditions as the one she now faced.

Again and again, it's been demonstrated that contact between individuals from diverse backgrounds leads to novel solutions that outperform teams made up of similar, homogeneous backgrounds and become more effective over time in identifying problems and generating solution alternatives. Identifying problems and coming up with insightful and innovative solutions is exactly what global teams like Molinas need as they face dynamic emerging threats and external threats alike.

However, in addition to diversity, global teams must share commonalities about how they will work together. One member of Molinas's new group recalled that they functioned well because "[t]he team connected at the level of core values such as sincerity, authenticity, and being non-political and mission-oriented." They were able to work collaboratively and discuss differences of opinion that eventually led to viable solutions. Molinas could see that people were happy, growing into their new roles, and talking about their careers in a positive way: "We were able to send a few people to other countries for short-term assignments. Now they are taking on bigger roles."

Uniting the leadership team required an enormous amount of work. Molinas had deliberately chosen strong, strategic thinkers, eager to appreciate and contribute unique insights. "I'm happy I made the call," she said. "What it means, though, is that it's going to require extra effort to ensure that the new team functions as a cohesive unit." Only two members of her original eight remained. Her quality assurance director reflected on her experience with the new diverse team: "Is diversity good? Yes, definitely! Is it easy? That's another question. I've had to get used to working with different people from different countries. Having a diverse team is good for different perspectives, but our colleagues' business culture is different than ours. In Turkey, in general, we have a culture of being nice to each other. We don't have an aggressive style. In working with the bottlers, managing the relationship is a key priority. The new members of the leadership team need some time to understand the culture here, the realities, and how we work."

The quality assurance director brings up an important point about working with diverse teams, namely, that individual comfort levels with confrontation and productive conflict differ according to culture as well as temperament. To help those who were reluctant to engage in open disagreement, she encouraged people to get to know one another on a personal level first. Once they felt familiar with one another, a sense of mutual trust would make it easier to disagree without fear of causing offense. She also found it useful to hire an outside consultant to work with two people who came from a military background and were used to a command-and-control style of decision-making instead of a collaborative, iterative process.

Whereas the previous team's relative sameness had created dynamics of cognitive agreement and efficiency, the new team was eventually able to create a healthy team dynamic of open debate, discussion, and friction—the qualities that ultimately led to innovative solutions. Molinas likened her team to the United Nations because of its broad array of perspectives, and laughed when she reported, "I believe I am seeing a much healthier business and healthier team dynamic. To that end, they are arguing more. And I love it."

ACT!

Molinas credited the external shocks of crisis with forcing her to act on what she already sensed about the limitations of leading a team that lacks diversity. She said it also made her realize the importance of acting immediately to prepare for the next crisis. In her words,

What I have learned is the moment you realize that something is not working, you have to go down to the last atomic piece of it to make

sure that you fix it. Otherwise things happen one after another and you can waste a lot of time. This is why, for the health of the organization, for the health of the business, you have to act the moment you see something. The moment you sense it.

In many cases, what needs to be addressed is the team composition. A diversity of cognitive approaches, which comes from a talent pool that is broadly representative of nationalities, societies, cultures, religions, racial backgrounds, and so forth, plays a crucial role in determining global teams' ability to adapt effectively to crises. Adaptation is a continuous quest to stay relevant in changing business environments. Diversity inspires the creative thinking necessary to undergo adaptation and stay relevant in ever-changing markets. Team members with diverse experiences across international markets themselves have already mastered the art of adaptation, adjusting themselves to novel working environments and vastly different markets, political contexts, or other scenarios.

Molinas found that eventually, insights flowed from the new leadership team members as they shared their background experiences, particularly within multiple emerging markets, and applied them as appropriate. For example, team members with experience in the Russian market led to insights appropriate for Central Asia. A team member with experience in the Venezuelan market, with its history of political upheaval and coups, led to transferable insights for Turkey. Similarly, Venezuela had a closed economy, with many parallels to Molinas's market in Uzbekistan.

The new leadership team also allowed her to move to a fully centralized structure, which she had resisted pre-crisis. She created a new general manager role for Turkey who became a member of the leadership team, along with a similar general manager in Central Asia.

Both managers were able to remain close to operations in their own markets and act quickly on insights gleaned.

Combined relevant experiences across markets allow leaders and teams to act accordingly with what the group learns relative to their markets. Diverse skills and backgrounds provide the cognitive diversity necessary to come up with ideas for building new business models. When diversity is centered on functional skills, it can aid in building new growth categories or additions to the core businesses in times of economic turbulence. Diverse teams require extra effort to ensure that teams function cohesively, but the insights the teams afford can solve for present challenges while opening potential opportunities.

While the composition of teams and the role of leaders are essential, the structure in which teams operate is just as important. That is, decisions about whether the team operates in a centralized, decentralized, or hybrid structure in a given geography will also influence process. A more centralized strategy can more readily identify existing and potential synergies that can help with decision-making and execution. For global teams, remaining close to operating teams to gain local market insights and act on them quickly is imperative.

EPILOGUE: MOLINAS IN MEXICO DURING COVID-19

I caught up with Molinas in July 2020. Although the number of newly infected people had then fallen to low levels in Massachusetts, a state that had been hit hard in the first months of the pandemic, and my neighborhood shops were beginning to reopen, the number of reported cases in the United States continued to rise. Molinas was now president of the Mexico business unit of the Coca-Cola Company. Mexico was also seeing growing number of cases and deaths. I wanted to know how Molinas was handling the enormous shocks we were all experiencing. Had the crisis in Istanbul prepared her for a global shutdown? How was she personally responding? What measures had she taken with her employees and colleagues?

Mexico Coca-Cola sent their workers home to work remotely on March 17. Molinas and her colleagues first framed the situation by spending a lot of time talking to anthropologists and sociologists, trying to understand what people were experiencing on an emotional and behavioral level, and trying to anticipate which changes might be transitory and which ones permanent. They also tried to anticipate how political leaders might respond, both in Mexico and around the world.

However, this first situation framing could only do so much. Molinas admitted that she prefers not to spend all her time trying to pin down what the future will hold, an activity she calls "fortune-telling." Instead, she found it more productive to simply accept that the global pandemic had "disrupted everything," and that no one—CEO or entry-level assistant—can be certain of an assessment. In place of certainty, she pursued clarity about what choices could be made and their impact on daily work.

To adapt to the new situation, she held virtual town-hall meetings where management addressed employee questions, organizational psychologists spoke about mental health challenges, and senior leaders offered insights. These thirty-minute daily meetings began the day after the company shut down and continued for fifty-eight days. Molinas felt that the communications conveyed in the daily town halls were especially necessary in a highly interactive culture where, pre-COVID, it had taken her at least twenty minutes to reach her office from the building entrance because of all the customary greetings, hugs, and inquiries about people's families. Ultimately, the town halls became, she said, "a dialogue and learning platform for all."

Molinas and her colleagues acted immediately by developing five principles designed to help manage what became the first one hundred days of the crisis; the first principle was, "People first with empathy at the core," and included "Manage today and emerge stronger" and "Ensure one voice as a system." As conditions changed, the principles evolved for ones best suited to manage the second hundred days. Her business unit streamlined the tasks by 50 percent, prioritized sixteen critical projects to manage today in order to emerge stronger, and aligned the company on critical business-essential tasks. They centralized allocation of the budget and human resources by establishing a transformation office in the middle of the crisis, in April. She also changed half of her leadership team. The team introduced three weekly routines: a decision-making forum, an investment committee, and coaching sessions for the project leads. They also focused and aligned resources for the year and developed a strategy with critical priorities to pursue diligently and deliver on.

Molinas told me that although her first response was to treat the public health emergency as any other crisis, one she could meet "thanks to the muscles and scars" that she'd acquired after having dealt with many past crises, her "aha" moment was when she realized, with a shock, that what was happening was much "bigger and more complex" than any crisis she'd encountered in the past. "This experience has been precious for me and it has influenced me in a profound way, as an individual," she said. At first I was struck by her use of the word *precious*, which usually pertains to something valuable or rare. But precious can also refer to something singular or unique, as in the last lines of Mary Oliver's well-known poem, "The Summer Day," when she asks what each of us plans to do with our "one wild and precious life." The more I thought about it the word choice seemed apt; if we can meet a crisis with eyes-wide-open awareness, learn to frame it as intelligently as possible, and then act as quickly and best as we know how, then we and our teams will have a profound and precious experience. One of the meanings of "revolution" is of one object's circular movement around another object. In the remote revolution, how we turn around one another and what we choose to do with our precious experience is up to each of us.

Success from Anywhere: Preparing for Global Crises

- **Scan current global issues** as the first step in developing panoramic awareness. Consume a variety of international media on a consistent basis to help anticipate how global issues may impact your local organization.
- **Frame the situation and risks** your team may face as you prepare to meet possible future challenges wrought by global events. Don't react defensively; instead, frame your situation intelligently and from a number of different close-up portrait angles to consider potential solutions.
- **Talk to colleagues, workers, and subject experts** to gain insight into how to best meet an ongoing crisis or to prepare for future crises.
- **Act immediately** as best you know how in response a crisis once you have reached a satisfactory strategy.
- **Prepare for radical change that may be required.** Strategize to meet the crisis and understand that actions may require deep structural reorganization, resource reallocations, leadership realignments, or other radical change.

Action Guide

This Action Guide is intended to help you and your team apply insights and best practices from each chapter to your own working environment. Each set of action practices is designed to take you deeper into the chapter content to enable reflection, learning, and application. Your remote team and leader will be asked to answer questions to sharpen your remote work acumen so that you can launch, trust, enhance productivity, use digital tools effectively, become more agile, work across differences, lead virtually, and prepare for global crises. The questions and exercises are also intended to foster bonding experiences as you share and discuss the material in *Remote Work Revolution* and the specific ways in which it applies to your team. The questions are not meant to be a test of your abilities! They are meant to help you succeed from anywhere.

You can use the Action Guide in a number of ways. You might want to engage in the action exercises immediately following the reading of each chapter in order to absorb the information and make it "stick." You might want to reach for the action exercise that seems most relevant to your immediate situation. Some leaders may send individual action exercises out to each team member to complete in preparation for an all-member meeting that takes place over the digital medium of your choice. Others may want to post the practice questions in an online collaborative tool that people can contribute to asynchronously and/or anonymously. Remember also that the action exercises can be revisited and repeated over time as team conditions change.

CHAPTER 1

How Can We (Re)Launch to Thrive in Remote Work?

The following practices will take your remote team through the first and most fundamental step in the remote work revolution: the team launch session. Treat these prompts as guideposts—or more precisely, launchpads—to frame your session and get you off the ground. Your team should cover the key areas of a successful launch session: refining the team's shared goals, establishing its communication norms, understanding each team member's contributions and constraints, and identifying the resources necessary for success. If you are a leader, you should convey your commitment to help the team thrive.

These actions should be repeated and adapted for each relaunch session as well. As the chapter made clear, launch sessions cannot be successful if they are treated as an isolated event that kicks off the journey before being abandoned. Launching and relaunching is a continuous process throughout the full duration of any team's life cycle—especially remote teams.

1. Describe your team's shared goals.

2. How would you describe your communication norms?

3. In the table below, record your thoughts on a relaunch discussion to improve your current communication norms.

Communication Norm	Impact

4. In the table below, create a list of your team members' contributions and constraints.

Team Member	Contributions	Constraints
Jenny	Jenny is a twenty-year company veteran and has a great deal of institutional knowledge.	She works remotely in a different time zone than the majority of the team.

5.   In the columns below, create a list of resources: what you need to meet team goals, how they will help you succeed, and where to locate them.

What	How	Where

6.   If you are a team leader, describe three ideas of how you can show commitment to your team through launch and relaunch sessions.

CHAPTER 2

How Can I Trust Colleagues I Barely See in Person?

The following action plan engages you and your team with key concepts for building trust among virtual team members: the trusting curve, cognitive passable trust, cognitive swift trust, emotional trust, direct knowledge, and reflected knowledge. The type of trust—and how much—varies based on a remote team's unique situation. These exercises help you determine what trust should look like in your own remote team—in relationships among team members and clients alike.

1.   **How can the trusting curve help your team determine the level of trust that is necessary to reach its goals? Please be specific.**

2.   **What is the difference between swift trust and passable trust? Use examples from your remote work to explain.**

3.   **Describe a person you have developed emotional trust with remotely in the last six months. What words or actions do you notice about that trusting relationship?**

4.   **Develop a plan that can help you gain direct knowledge in your relationships with virtual teammates, leading to a better understanding of their personal characteristics and behavioral norms.**

5. Develop a plan that can help you gain reflected knowledge in your relationships with virtual teammates, leading to deeper insight about how teammates see you and deeper empathy for their perspective.

6. Generate three ideas that can help you develop cognitive and emotional trust with virtual clients.

CHAPTER 3

Can My Team Really Be Productive Remotely?

There are three tried-and-true criteria for productivity that *thrives* on remote teamwork instead of subduing it: 1) delivering results; 2) facilitating individual growth; and 3) building team cohesion. On the team level, the following action practices will help you accurately assess your team's productivity, identify potential blind spots, and increase your team's cohesion. On the individual level, the exercises will help you elevate the contributions of teammates while enhancing your own remote work performance as well.

1. Assess your team's output to date. (See Sample Response)

Results	Met Expectations? (Yes or No)	Exceeded Expectations (Yes or No)	Explain
New Web App Tool	Yes	Yes	We met the client's basic needs to share project data, and then also created a dynamic user-friendly interface and added natural language processing functionalities, to go the extra mile.
Sales Goals	No	No	Goals were 16 percent under target

2. **How can remote work enhance your individual growth on the team?**

3. **Evaluate your team's cohesion. Describe any changes you have observed over time, and list potential next steps. (Sample Response)**

Evidence of Team Cohesion	Impact on Productivity	Next Steps
We doubled the amount of small group virtual meetings.	There seems to be less tension among virtual team members; they seem more connected.	We are planning to require daily virtual check-ins to see if it helps team cohesion even further and reassess in one month.

4. **What can you do to make teammates feel included on your remote team? Please be specific.**

5. **Comment on the list of attributes that describe your home conditions as a remote worker, and evaluate how each attribute impacts your job satisfaction and productivity.**

CHAPTER 4

How Should I Use Digital Tools in Remote Work?

Digital tools lay the infrastructure for remote teamwork. Without them, communication would not just be more difficult—it would be impossible. But as this chapter showed, not all digital tools are created equal. Different occasions call for different mediums. The following action practices prompt you to reflect on the key considerations for using digital tools most effectively on your team. On an individual level, the exercises will make you more precise in choosing the right digital tool for the right situation, and more communicative within each medium. On the team level, these practices will boost knowledge sharing among team members, making the team more collaborative as a whole.

1.   Describe the last time you experienced tech exhaustion. What would you do differently to avoid it in the future?

2.   How would you describe the primary differences between face-to-face interactions and digital communications?

3. Discuss the following numbered items with your team and decide the best digital tool in your organization for you to achieve each. For example, videoconferencing might be your choice for coordination that requires a synchronous and rich medium.

	Rich	Lean
Synchronous	1. Coordination 2. Discussion 3. Collaboration 4. Team building	7. Coordination 8. Information exchange
Asynchronous	5. Content development 6. Team selection	9. Content development 10. Information exchange 11. Simple coordination 12. Complex information

1.	7.
2.	8.
3.	9.
4.	10.
5.	11.
6.	12.

4. How well does your team share knowledge? How can you and your team improve?

5. **What do you consider to be the advantages and disadvantages of communicating using private social media tools with your team?**

CHAPTER 5

How Can My Agile Team Operate Remotely?

Impressive team wins within organizations of vastly different sizes and ages—from century-old multinational behemoths to digitally born tech startups—show the exciting synergy between agile methods and remote teams. The following action practices usher you down the aisle of this marriage in a series of overlapping steps: connecting your team with the underlying mission of agile methodologies, applying the agile method to your team's unique goals, reflecting more deeply on the ways that agile methods sync up with your team's remote format, and engaging more intentionally with the digital tools that facilitate this harmony. Each of these steps will help your team grasp the agile philosophy as a concept, and then execute it on the ground in demonstrable ways that are specific to your remote team's context.

1. **How can asynchronous communication tools help lubricate real-time discussions for your remote agile team?**

2.   **How can the agile method help your team?**

3.   **Describe how a remote format would improve your agile team's process. Provide at least two concrete examples.**

4.   **How can you provide stakeholders a better experience as a member of a remote agile team? Please be specific.**

CHAPTER 6

How Can My Global Team Succeed Across Differences?

The following action practices prompt you to reflect on the specific ways that you and your team members are both alike and different, how these differences may have been challenging in the past, and how to apply concrete norms to bridge these differences and build a shared team identity. On an individual level, each of these practices will reduce psychological distance among you and your team members. On an overall team level, the exercise of building stronger shared identity will make the group more cohesive and collaborative.

1. How can you help build ONE identity for your team?

2. Describe a time when you encountered unfamiliar beliefs or norms on your globally distributed team. What was that like?

3. Describe a time when you felt you had common ground with a team member from another cultural background. What was that like?

4. What would you like to learn from your team members? What can you teach them?

5. Think back on the last month and describe a difficult interaction you had with a native or a nonnative speaker on your team. Explain why the situation was challenging to you. Explain why the situation may have been challenging to them.

CHAPTER 7

What Do I Really Need to Know About Leading Virtually?

Meeting these challenges is a matter of adapting your in-person leadership tool kit into a virtual setting, and making deliberate efforts to lay the foundations of teamwork that may form more naturally on collocated teams. The following action practices will reinforce a remote leadership tool kit that prevents against the worst effects of faultlines (whether they be a result of status differentials, geographic dispersion, or cultural differences), maximizes the potential of each individual team member, and unites the team around its ultimate goals.

1. How would you describe the primary differences between leading a collocated team and leading virtually?

2. How do status differences manifest in your team? What are three things that you can do to minimize them?

3. How do you think your team would rate you on your communication presence? What should you do differently?

4. Using the table below, describe the individual strengths of your team members that can help you achieve your collective goal.

Team member	Strength

5. Identify and evaluate the faultlines that could negatively impact your team.

Faultline	Impact on team

CHAPTER 8

How Do I Prepare My Team for Global Crises?

The capacity to thrive within crisis depends on three skills: panoramic awareness, active anticipation, and acting immediately. The following action practices prompt you and colleagues to reflect on your team's unique position within the VUCA environment, and how each of the three skills can help your team respond directly to the inherent challenges. In each question, you and your team will apply general concepts from the chapter to the specific circumstances of your remote team.

1.   Describe the unique challenges that your team faces in a VUCA environment.

2.   How can the diversity of your teammates help your team face the challenges of a VUCA environment?

3.   How might you and your team be affected by the country-of-origin effect?

4.   How would you describe your team's preparedness for global crises?

5. Using the table below, evaluate your team's panoramic awareness,
active anticipation, and capacity to act immediately in the face of crisis.
When possible, provide specific examples to help explain your response.

Panoramic Awareness	Active Anticipation	Immediate Action

ACKNOWLEDGMENTS

I am fortunate that my work has always benefited from the contributions of a large and diverse community. Twenty years ago, I was convinced that technology would have a profound impact on the nature of work. This belief led me to do my doctoral training at Stanford University's Management Science and Engineering department with an amazing group focused on examining the intersection of work, technology, and organizations. I will forever be grateful to Steve Barley, Bob Sutton, Pam Hinds, and Diane Bailey for setting the foundation for a generation of scholars like me to examine how digital technology could facilitate work across boundaries.

Although remote and global work have been steadily rising for decades, I never imagined that a cataclysmic pandemic would force their proliferation to this extent and at such speed. The scale and scope of remote work worldwide has made it imperative for countless employees and managers to collaborate across boundaries. Over the years, I have been blessed with many intellectual partners who have influenced the development of the concepts, frameworks, and best practices for this work. A special thanks to Amy Bernstein, Robin Ely, Frances Frei, Bill George, Linda Hill, Karim Lakhani, Paul Leonardi, Jay Lorsch, Nitin Nohria, Jeff Polzer, Lakshmi Ramarajan, and Kyle Yee, who have given me precious insights along the way. I also extend special thanks to John Paul Hagan, Karen Propp, JT Keller, and Patrick Sanguineti, who have contributed tremendously to the research and development of this book. I am also very thankful to the Harvard Business School for generously providing significant resources to produce it.

I have always benefited from parents whose wisdom is only rivaled by their unconditional support. I am grateful for their insights, and encouragement. Words are not adequate to express my gratitude to my husband, Lawrence. I couldn't ask for a better emotional and intellectual partner. His extraordinary analytical mind coupled with his kind heart ensure that I am balanced in my own thinking. There is no one in the world that I'd rather be on lockdown with while writing about remote work remotely.

I am grateful to my editor at HarperCollins, Hollis Heimbouch, who understood the substance and the spirit of this content immediately; and Julia Eagleton, my agent, who encouraged me to write this book at the right time.

I finally turn to the thousands of people who shared their experiences, insights, anxieties, hopes, and concerns about virtual or global work for nearly two decades. The only way to understand the phenomena in this book is through the lived experiences of those in the trenches. I am thankful for the many people who entrusted me with their stories. My deepest hope is that this book will do justice to their generous contributions, and help everyone involved in remote work succeed from anywhere.

NOTES

Chapter 1: How Can We (Re)Launch to Thrive in Remote Work?

3   Hackman concluded: J. Richard Hackman, *Collaborative Intelligence: Using Teams to Solve Hard Problems* (Oakland: Berrett-Koehler, 2011), 155.

4   the team launch is what "breathes life" into the team: Ruth Wageman, Colin M. Fisher, and J. Richard Hackman, "Leading Teams When the Time Is Right: Finding the Best Moments to Act," *Organizational Dynamics* 38, no. 3 (2009): 194.

4   the four essential elements of teamwork: Wageman, Fisher, and Hackman, "Leading Teams," 193–203.

5   the fundamental goal of a launch session is alignment: Wageman, Fisher, and Hackman, "Leading Teams."

7   Remote team members often belong: John Mathieu, M. Travis Maynard, Tammy Rapp, and Lucy Gilson, "Team Effectiveness 1997–2007: A Review of Recent Advancements and a Glimpse Into the Future," *Journal of Management* 34, no. 3 (2008): 410–76.

7   Multiple team membership: Michael B. O'Leary, Anita W. Woolley, and Mark Mortensen, "Multiteam Membership in Relation to Multiteam Systems," in *Multiteam Systems: An Organization Form for Dynamic and Complex Environments*, ed. Stephen J. Zaccaro, Michelle A. Marks, and Leslie DeChurch (New York: Routledge, 2012), 141–72.

7   it is not uncommon: Mark Mortensen and Martine R. Haas, "Perspective—Rethinking Teams: From Bounded Membership to Dynamic Participation," *Organization Science* 29, no. 2 (2018): 341–55.

12   people continue to work through relevant topics: Alex Pentland, "The New Science of Building Great Teams," *Harvard Business Review* 90 (April 2012): 60–69.

13   Collocated teams tend to argue more: Mark Mortensen and Pamela J. Hinds, "Conflict and Shared Identity in Geographically Distributed Teams," *International Journal of Conflict Management* 12, no. 3 (2001): 212–38.

13   if psychological safety is not present: Amy C. Edmondson, *The Fearless*

Organization: Creating Psychological Safety in the Workplace for Learning, Innovation, and Growth (Hoboken, NJ: John Wiley & Sons, 2019).

14   remote workers' feelings of professional isolation lead: Timothy D. Golden, John F. Veiga, and Richard N. Dino, "The Impact of Professional Isolation on Teleworker Job Performance and Turnover Intentions: Does Time Spent Teleworking, Interacting Face-to-Face, or Having Access to Communication-Enhancing Technology Matter?," *Journal of Applied Psychology* 93, no. 6 (2008): 1412–21.

Chapter 2: How Can I Trust Colleagues I Barely See in Person?

20   Social scientists define trust: Daniel J. McAllister, "Affect- and Cognition-Based Trust as Foundations for Interpersonal Cooperation in Organizations," *Academy of Management Journal* 38, no. 1 (1995): 24–59.

21   How do we develop concerns for coworkers' welfare: Roy Y. J. Chua, Michael W. Morris, and Shira Mor, "Collaborating Across Cultures: Cultural Metacognition and Affect-Based Trust in Creative Collaboration," *Organizational Behavior Human Decision Processes* 118, no. 2 (2012): 116–31.

22   *passable trust*: Tsedal Neeley and Paul M. Leonardi, "Enacting Knowledge Strategy Through Social Media: Passable Trust and the Paradox of Non-Work Interactions," *Strategy Management Journal* 39, no. 3 (2018): 922–46.

22   social scientists have construed what's called *swift trust*: Brad C. Crisp and Sirkka L. Jarvenpaa, "Swift Trust in Global Virtual Teams: Trusting Beliefs and Normative Actions," *Journal of Personnel Psychology* 12, no. 1 (2013): 45.

22   When swift trust is the norm: Crisp and Jarvenpaa, "Swift Trust," 45–56.

24   In teams, trust includes an expectation: P. Christopher Earley and Cristina B. Gibson, *Multinational Work Teams: A New Perspective* (Mahwah, NJ: Lawrence Erlbaum, 2002).

24   emotional trust is grounded in: Daniel J. McAllister, "Affect- and Cognition-Based Trust as Foundations for Interpersonal Cooperation in Organizations," *Academy of Management Journal* 38, no. 1 (1995): 24–59.

25   Cognitive Trusting Curves: Mijnd Huijser, *The Cultural Advantage: A New Model for Succeeding with Global Teams* (Boston: Intercultural Press, 2006).

28   he felt a deep sense of shared accomplishment: This is a composite example adapted from a series of descriptive case studies on trust in global virtual teams: Sirkka L. Jarvenpaa and Dorothy E. Leidner, "Communication and Trust in Global Virtual Teams," *Organization Science* 10, no. 6 (1999): 791–815. This paper is the earliest and most cited study on the concept of swift trust in virtual teams.

29   can be most challenging for people: Norhayati Zakaria and Shafiz Affendi
Mohd Yusof, "Can We Count on You at a Distance? The Impact of Culture
on Formation of Swift Trust Within Global Virtual Teams," in *Leading
Global Teams: Translating Multidisciplinary Science to Practice*, eds. Jessica L.
Wildman and Richard L. Griffith (New York: Springer, 2015), 253–68.

29   Swift trust occurs in teams: Debra Meyerson, Karl E. Weick, and Roderick
M. Kramer, "Swift Trust and Temporary Groups," in *Trust in Organizations*,
eds. Roderick M. Kramer and Tom R. Tyler (Thousand Oaks, CA: Sage,
1996), 166–95.

29   Transparency, or sharing information freely: D. Sandy Staples and Jane
Webster, "Exploring the Effects of Trust, Task Interdependence and
Virtualness on Knowledge Sharing in Teams," *Info Systems Journal* 18, no. 6
(2008): 617–40.

29   in the beginning of a geographically distributed: Lucia Schellwies,
Multicultural Team Effectiveness: Emotional Intelligence as a Success Factor
(Hamburg: Anchor Academic Publishing, 2015).

32   Understanding the norms of one's own site: Mark Mortensen and Tsedal
Neeley, "Reflected Knowledge and Trust in Global Collaboration,"
Management Science 58, no. 12 (2012): 2207–24.

33   Here are elements of self-disclosure that matter to receivers: Paul C. Cozby,
"Self-Disclosure: A Literature Review," *Psychological Bulletin* 79, no. 2
(1973): 73–91; Valerian J. Derlega, Barbara A. Winstead, and Kathryn
Greene, "Self-Disclosure and Starting a Close Relationship," in *Handbook of
Relationship Initiation*, eds. Susan Sprecher, Amy Wenzel, and John Harvey
(New York: Psychology Press, 2008), 153–74; Kathryn Greene, Valerian J.
Derlega, and Alicia Mathews, "Self-Disclosure in Personal Relationships,"
in *The Cambridge Handbook of Personal Relationships*, eds. Anita L. Vangelisti
and Daniel Perlman (Boston: Cambridge University Press, 2006), 409–27.

Chapter 3: Can My Team Really Be Productive Remotely?

41   in addition to the software, she was to download: Bobby Allyn, "Your Boss
Is Watching You: Work-From-Home Boom Leads to More Surveillance,"
NPR: All Things Considered (blog), May 13, 2020, https://www.npr.org
/2020/05/13/854014403/your-boss-is-watching-you-work-from-home
-boom-leads-to-more-surveillance.

42   he promptly installed digital monitoring devices: Chip Cutter, Te-Ping
Chen, and Sarah Krouse, "You're Working from Home, but Your Company
Is Still Watching You," *Wall Street Journal*, April 18, 2020, https://www

.wsj.com/articles/youre-working-from-home-but-your-company-is-still-watching-you-11587202201?mod=searchresults&page=2&pos=18.

42   employees became highly stressed and felt disempowered: Clive Thompson, "What If Working from Home Goes on . . . Forever?," *New York Times*, June 9, 2020, https://www.nytimes.com/interactive/2020/06/09/magazine/remote-work-covid.html.

42   millennials had intentions to leave companies: "The Deloitte Global Millennial Survey 2020," Deloitte, June 2020, https://www2.deloitte.com/global/en/pages/about-deloitte/articles/millennialsurvey.html#infographic.

44   three criteria for establishing successful outcomes for teams: J. Richard Hackman, *Leading Teams: Setting the Stage for Great Performances* (Boston: Harvard Business School Press, 2002).

46   Cisco credited an uptick: *Work-Life Balance and the Economics of Workplace Flexibility*, prepared by the Council of Economic Advisers (Obama Administration), Executive Office of the President (Washington, D.C., March 2010), https://obamawhitehouse.archives.gov/files/documents/100331-cea-economics-workplace-flexibility.pdf.

46   Sun Microsystems: Tsedal Neeley and Thomas J. DeLong, *Managing a Global Team: Greg James at Sun Microsystems Inc. (A)*. Harvard Business School Case No. 409-003 (Boston: Harvard Business School Publishing, July 2008).

48   A group of economists teamed up: Nicholas Bloom, James Liang, John Roberts, and Zhichun Jenny Ying, "Does Working from Home Work? Evidence from a Chinese Experiment," *Quarterly Journal of Economics* 130, no. 1 (2015): 165–218.

50   they found a 4.4 percent increase: Prithwiraj (Raj) Choudhury, Cirrus Foroughi, and Barbara Larson, "Work-from-Anywhere: The Productivity Effects of Geographic Flexibility," *Academy of Management Proceedings*, (2020, forthcoming): 1–43.

51   Employees were surveyed: Donna Weaver McCloskey, "Telecommuting Experiences and Outcomes: Myths and Realities," in *Telecommuting and Virtual Offices: Issues and Opportunities*, ed. Nancy J. Johnson (Hershey, PA: Idea Group, 2011), 231–46.

52   feelings of exhaustion weakened these results: Timothy D. Golden, "Avoiding Depletion in Virtual Work: Telework and the Intervening Impact of Work Exhaustion on Commitment and Turnover Intentions," *Journal of Vocational Behavior* 69, no. 1 (2006): 176–87.

52   the employees who perceived: Ellen Ernst Kossek, Brenda A. Lautsch, and Susan C. Eaton, "Telecommuting, Control, and Boundary Management:

Correlates of Policy Use and Practice, Job Control, and Work-Family Effectiveness," *Journal of Vocational Behavior* 68, no. 2 (2006): 347–67.

55   the flexibility in timing and increased opportunity: David G. Allen, Robert W. Renn, and Rodger W. Griffeth, "The Impact of Telecommuting Design on Social Systems, Self-Regulation, and Role Boundaries," *Research in Personnel and Human Resources Management* 22 (2003): 125–63.

56   workers can collaborate productively on a remote team: Stefanie K. Johnson, Kenneth Bettenhausen, and Ellie Gibbons, "Realities of Working in Virtual Teams: Affective and Attitudinal Outcomes of Using Computer-Mediated Communication," *Small Group Research* 40, no. 6 (2009): 623–49.

56   "professional isolation": Timothy D. Golden, John F. Veiga, and Richard N. Dino, "The Impact of Professional Isolation on Teleworker Job Performance and Turnover Intentions: Does Time Spent Teleworking, Interacting Face-to-Face, or Having Access to Communication-Enhancing Technology Matter?," *Journal of Applied Psychology* 93, no. 6 (2008): 1416.

56   research has identified loneliness: Nick Tate, "Loneliness Rivals Obesity, Smoking as Health Risk," WebMD, May 4, 2018, https://www.webmd.com /balance/news/20180504/loneliness-rivals-obesity-smoking-as-health-risk.

57   the researchers found no negative correlations: Timothy D. Golden and Ravi S. Gajendran, "Unpacking the Role of a Telecommuter's Job in Their Performance: Examining Job Complexity, Problem Solving, Interdependence, and Social Support," *Journal of Business and Psychology* 34 (2019): 55–69.

57   For some job features, performance: Cynthia Corzo, "Telecommuting Positively Impacts Job Performance, FIU Business Study Reveals," *BizNews .FIU.Edu* (blog), February 20, 2019, https://biznews.fiu.edu/2019/02 /telecommuting-positively-impacts-job-performance-fiu-business-study -reveals/.

58   employees did better on creative: Ronald P. Vega and Amanda J. Anderson, "A Within-Person Examination of the Effects of Telework," *Journal of Business and Psychology* 30 (2015): 319.

Chapter 4: How Should I Use Digital Tools in Remote Work?

61   Breton had been contemplating: Tsedal Neeley, J. T. Keller, and James Barnett, *From Globalization to Dual Digital Transformation: CEO Thierry Breton Leading Atos Into "Digital Shockwaves" (A)*. Harvard Business School Case No. 419-027 (Boston: Harvard Business School Publishing, April 2019).

61   "We are producing data on a massive scale": David Burkus, "Why Banning Email Works (Even When It Doesn't)," *Inc.*, July 26, 2017, https://www.inc .com/david-burkus/why-you-should-outlaw-email-even-if-you-dont-succe .html.

62   Internal emails were replaced: Max Colchester and Geraldine Amiel, "The IT Boss Who Shuns Email," *Wall Street Journal,* November 28, 2011, https://www.wsj.com/articles/SB100014240529702044521045770601031653 99154.

62   his bold plan dramatically decreased: Burkus, "Banning Email."

66   Why do remote work goals: Catherine Durnell Cramton, "The Mutual Knowledge Problem and Its Consequences for Dispersed Collaboration," *Organization Science* 12, no. 3 (2001), 346–71.

68   what social scientists call *social presence*: John Short, Ederyn Williams, and Bruce Christie, *The Social Psychology of Telecommunications* (London: Wiley, 1976).

70   they will start talking about rich or lean media: Richard L. Daft and Robert H. Lengel, "Organizational Information Requirements, Media Richness, and Structural Design," *Management Science* 32, no. 5 (1986): 554–71.

70   communication is made up of two primary processes: Alan R. Dennis, Robert M. Fuller, and Joseph S. Valacich, "Media, Tasks, and Communication Processes: A Theory of Media Synchronicity," *MIS Quarterly* 32, no. 3 (2008): 575–600.

71   Other researchers took this work a step further: Jolanta Aritz, Robyn Walker, and Peter W. Cardon, "Media Use in Virtual Teams of Varying Levels of Coordination," *Business and Professional Communication Quarterly* 81, no. 2 (2018): 222–43; Dennis, Fuller, and Valacich, "Media, Tasks."

73   negotiations and group decision-making: Roderick I. Swaab, Adam D. Galinsky, Victoria Medvec, and Daniel A. Diermeier, "The Communication Orientation Model Explaining the Diverse Effects of Sight, Sound, and Synchronicity on Negotiation and Group Decision-Making Outcomes," *Personality and Social Psychology Review* 16, no. 1 (2012): 25–53.

73   consider the dynamics and history of your teams: Swaab et al., "Communication Orientation Model."

74   technologies that increase presence awareness: Arvind Malhotra and Ann Majchrzak, "Enhancing Performance of Geographically Distributed Teams Through Targeted Use of Information and Communication Technologies," *Human Relations* 67, no. 4 (2014): 389–411.

74   you've almost certainly been: Paul M. Leonardi, Tsedal B. Neeley, and Elizabeth M. Gerber, "How Managers Use Multiple Media: Discrepant

Events, Power, and Timing in Redundant Communication," *Organization Science* 23, no. 1 (2012): 98–117. For communication to be considered redundant, the message had to contain the same general information as the initial communication. It could not contain new information, and it could not ask the receiver to engage in any new activity. In other words, the message could not introduce anything new even if the language used a second time around differed from the original by phrasing such as "as I mentioned before" or "please remember" that indexed the second communication to the communication that came before. Our general rule of thumb was that a communication was redundant if the message contained roughly 80 percent of the same information from the first message. If this quantity of information was the same, we coded both instances of media use together as a redundant communication and applied one code to it. For example, if we found that a manager used the telephone to call a team member to give him some figures to include in a report, and later that informant sent those same figures to the same team member via email, we coded the entire redundant communication episode as Phone→email.

78 enhanced trust and helped bolster a sense of team identity: Pnina Shachaf, "Cultural Diversity and Information and Communication Technology Impacts on Global Virtual Teams: An Exploratory Study," *Information & Management* 45, no. 2 (2008): 131–42.

78 Leaner communication technologies allowed: Anders Klitmøller and Jakob Lauring, "When Global Virtual Teams Share Knowledge: Media Richness, Cultural Difference and Language Commonality," *Journal of World Business* 48, no. 3 (2013): 398–406.

78 What might be regarded as common: Norhayati Zakaria and Asmat Nizam Abdul Talib, "What Did You Say? A Cross-Cultural Analysis of the Distributive Communicative Behaviors of Global Virtual Teams," 2011 International Conference on Computational Aspects of Social Networks (CASoN) (2011): 7–12.

79 Our modern lives are defined: Tsedal B. Neeley and Paul M. Leonardi, "Enacting Knowledge Strategy through Social Media: Passable Trust and the Paradox of Non-Work Interactions," *Strategic Management Journal* (in press).

Chapter 5: How Can My Agile Team Operate Remotely?

86 a new method for software development: Kent Beck, Mike Beedle, Arie van Bennekum, Alistair Cockburn, et al., "Manifesto for Agile Software Development," 2001, https://agilemanifesto.org/.

86 "We gave up on trench warfare": Jeff Sutherland and J. J. Sutherland, *Scrum: The Art of Doing Twice the Work in Half the Time* (New York: Crown, 2014), 6.

88 comfortable with a high degree of autonomy and accountability: Stephen Denning, *The Age of Agile: How Smart Companies Are Transforming the Way Work Gets Done* (New York: Amacom, 2018).

89 "The most efficient and effective": Beck et al., "Manifesto."

89 face-to-face communication makes teams: Subhas Misra, Vinod Kumar, Uma Kumar, Kamel Fantazy, and Mahmud Akhter, "Agile Software Development Practices: Evolution, Principles, and Criticisms," *International Journal of Quality & Reliability Management* 29, no. 9 (2012): 972–80.

89 Face-to-face conversation has been seen: Sutherland and Sutherland, *Scrum*.

90 LEGO created: Alesia Krush, "5 Success Stories That Will Make You Believe in Scaled Agile," *ObjectStyle* (blog), January 13, 2018, https://www.objectstyle.com/agile/scaled-agile-success-story-lessons.

90 the team accomplished more work for 3M: Paul LaBrec and Ryan Butterfield, "Using Agile Methods in Research," *Inside Angle* (blog), 3M Health Information Systems, June 28, 2016, https://www.3mhisinsideangle.com/blog-post/using-agile-methods-in-research/.

91 NPR was able to achieve significant: Hrishikesh Bidwe, "4 Examples of Agile in Non-Technology Businesses," Synerzip, May 23, 2019, https://www.synerzip.com/blog/4-examples-of-agile-in-non-technology-businesses/.

91 The bank's Net Promoter Score: Andrea Fryrear, "Agile Marketing Examples & Case Studies," AgileSherpas, July 9, 2019, https://www.agilesherpas.com/agile-marketing-examples-case-studies/.

92 ING's implementation of agile: William R. Kerr, Federica Gabrieli, and Emer Moloney, *Transformation at ING (A): Agile*. Harvard Business School Case 818-077 (Boston: Harvard Business School Publishing, revised May 2018).

95 AppFolio was *born* digital: Tsedal Neeley, Paul Leonardi, and Michael Norris, *Eric Hawkins Leading Agile Teams @ Digitally-Born AppFolio (A)*. Harvard Business School Case 419-066 (Boston: Harvard Business School Publishing, revised February 2020).

Chapter 6: How Can My Global Team Succeed Across Differences?

111 Tariq Khan sat: Tsedal Neeley, *(Re)Building a Global Team: Tariq Khan at Tek*. Harvard Business School Case 414-059 (Boston: Harvard Business School Publishing, revised November 2015).

114 German sociologist Georg Simmel published: Georg Simmel, "The Stranger," in *The Sociology of Georg Simmel* (Glencoe, IL: Free Press, 1950), 402–8.

118 challenge facing teams: Tsedal Neeley, *The Language of Global Success: How a Common Tongue Transforms Multinational Organizations* (Princeton, NJ: Princeton University Press, 2017).

124 Rules of Engagement: Adapted from Tsedal Neeley, "Global Teams That Work," *Harvard Business Review* 93, no. 10 (2015), 74–81.

Chapter 7: What Do I Really Need to Know About Leading Virtually?

133 Leadership is empowering other people: Frances Frei and Anne Morriss, *Unleashed: The Unapologetic Leader's Guide to Empowering Everyone Around You* (Boston: Harvard Business School Press, 2020).

135 where people are located: For the purposes of this book, team structure refers to the physical configuration of teams. It is well acknowledged that in the teamwork literature, team structure encompasses a much broader range of attributes including task allocation, authority, roles and responsibilities, norms, and patterns of interaction, among others. See Greg L. Stewart and Murray R. Barrick, "Team Structure and Performance: Assessing the Mediating Role of Intrateam Process and the Moderating Role of Task Type," *Academy of Management Journal* 43, no. 2 (2000): 135–48; Daniel R. Ilgen, John R. Hollenbeck, Michael Johnson, and Dustin Jundt, "Teams in Organizations: From Input-Process-Output Models to IMOI Models," *Annual Review of Psychology* 56 (2005): 517–43.

135 Configuration describes: Michael Boyer O'Leary and Jonathon N. Cummings, "The Spatial, Temporal, and Configurational Characteristics of Geographic Dispersion in Teams," *MIS Quarterly* 31, no. 3 (2007): 433–52; Michael B. O'Leary and Mark Mortensen, "Go (Con)figure: Subgroups, Imbalance, and Isolates in Geographically Dispersed Teams," *Organization Science* 21, no. 1 (2010): 115–31.

136 members who sit: David J. Armstrong and Paul Cole, "Managing Distances and Differences in Geographically Distributed Work Groups," in *Distributed Work*, eds. Pamela Hinds and Sara Kiesler (Cambridge, MA: MIT Press, 2002), 167–86.

136 Solo workers can feel threatened: Jeffrey T. Polzer, C. Brad Crisp, Sirkka L. Jarvenpaa, and Jerry W. Kim, "Extending the Faultline Model to Geographically Dispersed Teams: How Colocated Subgroups Can Impair Group Functioning," *Academy of Management Journal* 49, no. 4 (2006): 679–92.

137   they were more likely to communicate: Paul M. Leonardi and Carlos
Rodriguez-Lluesma, "Occupational Stereotypes, Perceived Status
Differences, and Intercultural Communication in Global Organizations,"
Communication Monographs 80, no. 4 (2013): 478–502.

138   Researchers refer to faultlines: Dora C. Lau and J. Keith Murnighan,
"Demographic Diversity and Faultlines: The Compositional Dynamics of
Organizational Groups," *Academy of Management Review* 23, no. 2 (1998):
325–40.

139   Faultlines cause trouble when the cracks widen: Katerina Bezrukova, Karen
A. Jehn, Elaine L. Zanutto, and Sherry M. B. Thatcher, "Do Workgroup
Faultlines Help or Hurt? A Moderated Model of Faultlines, Team
Identification, and Group Performance," *Organization Science* 20, no. 1
(2009): 35–50.

140   My study with two colleagues: Pamela J. Hinds, Tsedal Neeley, and
Catherine Durnell Cramton, "Language as a Lightning Rod: Power
Contests, Emotion Regulation, and Subgroup Dynamics in Global Teams,"
Journal of International Business Studies 45, no. 5 (June–July 2014): 536–61.

141   Groups often thrive on faultlines: Bezrukova et al., "Workgroup Faultlines."

141   *group-level identity*: Naomi Ellemers, Dick De Gilder, and S. Alexander
Haslam, "Motivating Individuals and Groups at Work: A Social
Identity Perspective on Leadership and Group Performance," *Academy of
Management Review* 29, no. 3 (2004): 459–78.

143   Team members also respond favorably: Doreen B. Ilozor, Ben D. Ilozor,
and John Carr, "Management Communication Strategies Determine Job
Satisfaction in Telecommuting," *Journal of Management Development* 20,
no. 6 (2001): 495–507.

143   Researchers who wanted to know: Donna W. McCloskey and Magid
Igbaria, "Does 'Out of Sight' Mean 'Out of Mind'? An Empirical
Investigation of the Career Advancement Prospects of Telecommuters,"
Information Resources Management Journal 16, no. 2 (2003): 19–34.

144   a leader must attend to the traditional: Jeffrey Polzer, "Building Effective
One-on-One Work Relationships," Harvard Business School No. 497-028
(Boston: Harvard Business School Publishing, 2012).

Chapter 8: How Do I Prepare My Team for Global Crises?

153   a term used by the U.S. Army War College: Richard H. Mackey Sr.,
Translating Vision into Reality: The Role of the Strategic Leader (Carlisle
Barracks, PA: U.S. Army War College, 1992).

158   The country-of-origin effect: Robert D. Schooler, "Product Bias in the Central American Common Market," *Journal of Marketing Research* 2, no. 4 (1965): 394–97.

160   Mexican consumers began boycotting American products: Jack Jenkins, "Why Palestinians Are Boycotting Airbnb," ThinkProgress, January 22, 2016, https://archive.thinkprogress.org/why-palestinians-are-boycotting -airbnb-d53e9cf12579/; Ioan Grillo, "Mexicans Launch Boycotts of U.S. Companies in Fury at Donald Trump," *Time*, January 27, 2017, http://time .com/4651464/mexico-donald-trump-boycott-protests/.

160   heavily trending hashtags: Grillo, "Mexicans Launch Boycotts."

162   the federal government had not issued: David Montgomery, Ariana Eunjung Cha, and Richard A. Webster, "'We Were Not Given a Warning': New Orleans Mayor Says Federal Inaction Informed Mardi Gras Decision Ahead of Covid-19 Outbreak," *Washington Post*, March 27, 2020, https:// www.washingtonpost.com/national/coronavirus-new-orleans-mardi-gras /2020/03/26/8c8e23c8-6fbb-11ea-b148-c4cc3fbd85b5_story.html.

162   when leaders don't heed warning signals: Erika Hayes James and Lynn Perry Wooten, "Leadership as (Un)usual: How to Display Competence in Times of Crisis," *Organizational Dynamics* 34, no. 2 (2005): 141–52.

163   leadership at many levels: Li Yang Hsu and Min-Han Tan, "What Singapore Can Teach the U.S. About Responding to Covid-19," *Stat*, March 23, 2020, https://www.statnews.com/2020/03/23/singapore-teach-united -states-about-covid-19-response/.

168   A diversity of cognitive approaches: Katherine W. Phillips, Gregory B. Northcraft, and Margaret A. Neale, "Surface-Level Diversity and Decision-Making in Groups: When Does Deep-Level Similarity Help?," *Group Processes & Intergroup Relations* 9, no. 4 (2006): 467–82.

INDEX

Note: Italic page numbers refer to charts.